rr   versity
         55

FAST
Creativity
& Innovation

Rapidly Improving Processes, Product Development and Solving Complex Problems

Read the Reviews

"I have used FAST to improve an orthopedic procedure replacing a knee joint for the producer of prosthesis. The closing comment was that 'this is the first time the medical doctors and engineers were able to communicate on the same level.'"

> —*Jerry Kaufman, CVS, FSAVE*
> *President, J.J. Kaufman Associates, Inc.*

"The FAST approach breaks down function requirements into components and it represents the logical relationships between them. Working with functions helps to break through the common psychological inertia that so commonly stifles creativity in problem solving. On a technical level, FAST takes the logic associated with the structural paradigm and puts it in a form that is intuitive to use. Its primary use is to solve problems, but as a fundamental representation scheme new uses continue to emerge including a tool for teaching and a way to represent technology for research. This book will not only be of use to understand the method, but also to understand the original insight of its developer. As such, it is a landmark work in the field of creative problem solving."

> —*Dr. Martin Hyatt, Ph.D.*
> *Developer of Synoptics Creativity*

"We have undertaken plant layout studies, labor utilization, complete manufacture and packaging of panty hose, procurement systems, inventory control systems, corporate cash flow,...invoicing systems, debtors' systems, credit payment systems. We are confident that, if we keep to the rules of your FAST approach, the answer we seek will be found."

> —*Leon M. Turner, Management Consultant*
> *John P. Young & Associates*
> *Hawthorne, Australia*

"FAST reduces the time for complex audit and analysis of management planning to perhaps one-third of the time it would otherwise take, and at the same time increases the quality of the planning process and plan presentation."

> —*Donald P. Goss, Director*
> *Bureau of Systems Analysis*
> *Commonwealth of Pennsylvania*

"In this book, the author presents thought-provoking questions to enhance logic thinking and intuitive role-playing. These questions aid in the construction of function logic diagrams and offer additional insights. The function tree makes it possible for several people, working in different geographical locations, to work together with the use of a computer and e-mail, to develop the required functions for analysis and to generate creative solutions to problems."

—*Richard J. Park, PE, CVS, FSAVE*
President, RJ Park & Associates, Inc.

"To my knowledge, FAST is the only available technique for real product structural analysis."

—*C.A. Lopez Padilla*
Manager, Manufacturing Engineering
General Electric Argentina

"One Pennsylvania agency produced an agency plan by another method which required 638 pages. FAST would produce a more comprehensive and meaningful plan using less than 50 pages!"

—*John S. Hollar*
Cost Reduction Program Supervisor
Commonwealth of Pennsylvania

"We have applied the FAST Diagram technique to over 100 projects. We are also using the FAST approach in our application task forces to develop new designs, improve old products, analyze operational procedures, and invent new approaches. Results are extremely satisfactory. The FAST Diagram clarifies a problem and pinpoints the area to apply creativity."

—*Richard J. Park*
Manager, Value Control
Chrysler Corporation

FAST
Creativity
& Innovation
Rapidly Improving Processes,
Product Development and
Solving Complex Problems

Charles W. Bytheway

Copyright ©2007 by J. Ross Publishing, Inc.

ISBN-10: 1-932159-66-5
ISBN-13: 978-1-932159-66-0

Printed and bound in the U.S.A. Printed on acid-free paper
10 9 8 7 6 5 4 3 2 1

Library of Congress Cataloging-in-Publication Data

Bytheway, Charles W., 1926-
    FAST creativity & innovation : rapidly improving processes, product
development, and solving complex problems / Charles W. Bytheway.
      p. cm.
    Includes index.
    ISBN-10: 1-932159-66-5 (hardcover : alk. paper)
    ISBN-13: 978-1-932159-66-0
  1. Production engineering—Data processing. 2. Flow chart generators. 3.
Value analysis (Cost control)—Data processing. 4. New products. I. Title.
II. Title: Function analysis systems technique creativity and innovation.
    TS171.4.B98 2006
    658.5—dc22   2006025683

Phone: (954) 727-9333
Fax: (561) 892-0700
Web: www.jrosspub.com

TABLE OF CONTENTS

FOREWORD

Charles Bytheway is a giant when it comes to creative thinking, and we have much to thank him for. He discovered what I call the theory of function relationships and invented the FAST Diagram to illustrate the relationship. These two insights are major tools to aid in increasing a person's creative ability. They both have been incorporated in business methods around the world.

Here we have two great ideas, creativity and value analysis. The value analysis concept was developed by Lawrence D. Miles in the 1950s. The basis for the concept was that the way things look and work limits imagination to existing products and methods, but if we concentrate on what they do or what we want — the function — the result is unlimited creativity.

The value analysis management system is based on what is called function analysis. Function is defined as something that we want or need, a goal, objective, something we are willing to pay for, and functions must be defined in a specific way to foster creative development.

Charles studied this goal and discovered that every function has a cause and effect relationship, and he invented the FAST Diagram to illustrate the condition. In 1965 he presented a paper, "Basic Function Determination Technique," at the Society of American Value Engineers International Conference in Boston. In this paper he pointed out the function relationship and the FAST Diagram.

These methods have been accepted by value analysis specialists and engineers around the world, and in many cases it is believed that if you are not using FAST Diagrams, you are not doing value analysis. Needless to say, this has spawned a number of FAST variations, some good and some not so good. However, in almost every case it has been found that the effort expended in creating the diagram has fostered creative development. This book presents a FAST procedure that standardizes the method for creating FAST Diagrams and his function trees. The result of this work was that Charles is the first person to have been awarded the Miles Award, by the Society of American Value Engineers, for his work in advancing value analysis/engineering.

Recently, he has been asked by the society, now known as SAVE International, to put his thoughts on paper for everyone's benefit in understanding his thinking processes and to aid others in the use of his methods. He has done that in this book, and on the way he has expanded the function tree, which he introduced in his original SAVE Conference paper in 1965. The function tree makes it possible for several people, working in different geographical locations, to work together, with the use of a computer and e-mail, to develop the required functions for analysis and to generate creative solutions to problems. In this book, the author presents the basics of function analysis, plus his thought-provoking questions to enhance logic thinking and intuitive role-playing. In addition, these questions aid in the construction of function logic diagrams and offer additional insights into his thinking processes, which make this book a valuable text for creative classes as well for individual creative situations.

The function tree method, together with the FAST Diagram, is the most powerful mind-opening process for creative solutions available, and everyone who is engaged in work that requires creative solutions to problems will benefit from this book.

—Richard J. Park, PE, CVS, SAVE

FOREWORD

The field of creative problem solving is relatively new. There is much in the way of gimmicks or techniques that work in certain areas, but there is little in the way of substance. In my own research in the cognitive science of creative problem solving, two types of contributions emerged over the last 50 years. One kind of contribution is to draw from certain basic paradigms of creative problem solving. Some notable figures and their contributions are:

- **Genrikh Altshuller** draws from a structural paradigm in which the method TRIZ is developed.
- **James L. Adams** draws from a representational paradigm in his rules of perception.
- **Edward de Bono** draws from a state transformation paradigm in developing his lateral thinking.

There is another type of contribution that is just as important, which is to formalize these foundational ways of thinking into a representation and knowledge structure for the problem-solving process. Charles Bytheway made such a contribution with his Function Analysis System Technique, commonly referred to as FAST.

The FAST approach breaks down function requirements into components, and it represents the logical relationships among them. Working with functions helps to break through the common psychological inertia that so commonly stifles creativity in problem solving.

On a technical level, Bytheway's contribution was to take the logic associated with the structural paradigm and put it in the form that is intuitive to use. Its primary use is to solve problems, but as a fundamental representation scheme new uses continue to emerge, including a tool for teaching and a way to represent technology for research.

This logic of functions, requiring the combination of other functions, is called conjunctive logic. Bytheway's method was the first to use this logic.

Since then, there have been several rediscoveries of this same method that go under different names, such as function analysis, the Theory of Constraints, and the fishbone diagram. FAST preceded these other approaches by over 20 years. This book will be of use not only to understand the method but also to understand the original insight of its developer. As such, it is a landmark work in the field of creative problem solving.

—Dr. Martin Hyatt, Ph.D.

ABOUT THE AUTHOR

Charles Bytheway graduated from the University of Utah with a bachelor of science in mechanical engineering in 1952 and a master of science in mechanical engineering in 1961. Shortly before receiving the second degree, he received his value engineering certificate from the University of California at Los Angeles in 1960.

His interest in designing, building, and improving and his experience in carpentry during his university years led him to design and build two of his own homes. Before being hired by Sperry Rand Corporation, he designed heating and air-conditioning systems for homes and businesses for Lennox Industries, Inc. and designed underground mining equipment for EIMCO Corporation. While working for Sperry Rand, he designed a turret for a 62mm Gatling gun during the Vietnam War, worked on grenade launchers and missile systems, and conducted the first value engineering seminar within the corporation in 1960. He was the configuration manager for DDR, ARTS III, NAVAIR, the Navy's LAMPS program, and several other military programs.

Until retiring in 1981, Mr. Bytheway served as the director of value engineering in Salt Lake City. Not out of the value engineering game, he served as president of functional research, where he worked with a team that performed value engineering consulting on new construction for the University of California at Berkeley, Los Angeles, and San Diego. Then he went to work for the

Salt Lake Community College and started its mechanical engineering department, becoming an associate professor in that department.

During those years and following, he published 19 papers on FAST diagramming and related subjects, two of which were on Boolean logic. He has had several books published pertaining to scriptural quotations. He combined the four gospels of the New Testament into a single gospel text which was published in 1988.

Mr. Bytheway was the first recipient of the Lawrence D. Miles Award from the Society of American Value Engineers (now SAVE International) for his creative development of FAST diagramming. He is a Certified Value Specialist (CVS) and a Fellow of SAVE International. He was recommended to be on President Ronald Reagan's panel of experts to control runaway government spending, called the Private Sector Survey on Cost Control.

Value engineering and creativity do not, however, define the whole of Mr. Bytheway's life. A devout follower of the Church of Jesus Christ of Latter-Day Saints, he served a mission in New Zealand for 27 months and has served in bishoprics and other ecclesiastical positions of leadership within his church for the past 45 years. He is married to the former Frieda Duehlmeier and recently celebrated their 57th wedding anniversary. They have 4 daughters and 2 sons, 23 grandchildren, and 2 great granddaughters.

Mr. Bytheway can be contacted by visiting www.fastcreativity.com.

ACKNOWLEDGMENTS

Over 45 years ago, Ed Heller made an impression upon my life that I have never forgotten when he taught me about functions. I acknowledge him for putting that spark in my life that changed my professional career. Lawrence Miles, the founder of value analysis and value engineering, became a good friend and mentor over the years. He is the gentleman who changed the way we look at things.

I wish to recognize the great contribution which my friend Richard J. Park of Birmingham, Michigan has made toward the publication of this book. Without his guidance and editorial skills, my lifetime dream of writing a book about FAST never would have become a reality. Richard has always been a strong advocate of FAST and has been influential in teaching many others this technique. I also thank the many people who have written me over the years about their successes as they have used my technique.

During the past 30 years, Jerry Kaufman has sent me volumes of FAST Diagrams he and his associates have developed. Jerry's untiring effort in promoting FAST nationally and internationally is greatly appreciated. Thanks is extended to many others, too numerous to name, who have sent me materials to be used in my book and who have promoted the virtues of developing FAST Diagrams. Finally, I wish to thank my son Jay and my brother Alvin, who have made many constructive improvements in the text. I also wish to thank my wife, Frieda, for her patience.

—*Charles W. Bytheway*

**Web
Added
Value**

*Free value-added materials available from
the Download Resource Center at www.jrosspub.com*

At J. Ross Publishing we are committed to providing today's professional with practical, hands-on tools that enhance the learning experience and give readers an opportunity to apply what they have learned. That is why we offer free ancillary materials available for download on this book and all participating Web Added Value™ publications. These online resources may include interactive versions of material that appears in the book or supplemental templates, worksheets, models, plans, case studies, proposals, spreadsheets and assessment tools, among other things. Whenever you see the WAV™ symbol in any of our publications, it means bonus materials accompany the book and are available from the Web Added Value Download Resource Center at www.jrosspub.com.

Downloads available for *FAST Creativity & Innovation* include thought-provoking questions, ground rules for brainstorming, a FAST Tree for the FAST procedure, and templates for creating FAST Diagrams.

INTRODUCTION

Creative people look for opportunities to extend their imagination into areas where others have made assumptions or areas others have not considered or ventured into. Dr. Albert Einstein has been termed a conceptual inventor or genius. He said, "Imagination is more important than knowledge." He found that by exercising his imagination, he could extend his imagination into unknown areas. I believe that you too can venture into new creative areas if you apply the principles of creativity presented in this book.

UTILIZING ONE'S IMAGINATION

The creating of ideas or utilizing one's imagination is no longer the secret of just a few educated and successful men and women. It is available to you, and history proves it. A boy who worked in a meat market and sold candy, soda water, and magazines on a train in his spare time increased his ability to think up new ideas that brought success and fame. That boy was Thomas Edison. Great ideas have generally come from people who were working in unrelated fields of endeavor. Samuel F. Morse was a portrait painter; he invented the telegraph. The steamboat was invented by an artist, Robert Fulton. Eli Whitney, a schoolteacher, invented the cotton gin and was the first person to build parts that were interchangeable, which made clock making a thriving industry.

Their success was based on personal decisions to think more deeply about things they observed from day to day. We have a lot more things to observe than they did. How many times have you seen a new product come on the market and thought to yourself, "I could have come up with that idea if I had just taken the time to think about it." Those opportunities are everywhere, just waiting for you to recognize them.

Earl Tupper is a more recent inventor. He took black polyethylene slag, a waste product from oil refineries, and made a tough, flexible, nonporous, nongreasy, and translucent plastic, known as Tupperware®. An inventor by the name of Chester F. Carlson invented xerography and electrophotography. He is said to be the man who started it all, which has made it possible to print more than 100 copies a minute, to record and transmit electrostatic images and recordings, etc.

Every person has some creative ability. You are creative if you do one thing different today than you did yesterday. If you do two creative things tomorrow and add an additional one each day, your capabilities will continue to increase. My FAST Creativity technique will teach you how to stimulate your creativity and increase your creative opportunities. FAST is the acronym for Function Analysis System Technique. The logic questions involved in this technique are self-stimulating. Each answer is used to formulate two new questions. Both of these new questions force thinking into higher levels of understanding and into other methods of performing the same task.

For example, this technique will allow you to expand a simple one-line statement into a volume of information within a short period of time. As the information comes rolling in, it will spark your creativity and new ideas will begin to flow within your head. You have to experience it to believe it.

QUESTIONS TO ASK

This book teaches you the basic elements of function analysis and how FAST Creativity can change your method of thinking. It will teach you what questions to ask when you are selecting a task. Then it teaches you the questions to ask yourself, such as why you should devote energy toward a given task. Additional questions bring new facts to your attention and allow you to logically organize them. The logic associated with these questions also will help you identify any information that is missing.

Ten stimulating questions are used to analyze any subject. These questions will expand and enhance your thinking into new levels of understanding as this technique organizes your thinking and the information you have collected. When this happens, your creative mind will begin to ponder and apply those same questions to the new information you have gathered, and additional new ideas will begin to flow.

This path to creativity reminds me of when our grandchildren were living with us. My grandson was always asking me "why" over and over again about everything. As soon as I gave him a simple answer, he would ask "why" again and again about something else. As he got older, he started to ask "how" do you do this and "how" do you do that, over and over again. You see, as he got older, he was able to completely understand the reason "why" I was doing something; he then wanted to know "how" to do it himself.

WHY-HOW LOGIC

This proven "Why-How Logic" also taught you when you were young, and it continues to teach you today if you will take the time to recognize it. This Why-How Logic is the heart and meat of this creative technique. Maturity and experience help all of us to think deeper in so many different areas when we ask the same proven "why" and "how" questions. These two questions bring together facts so you can logically connect them and also understand them. They stimulate your creativity so you feel better about yourself as you experience an increase in your level of thinking and satisfaction in your accomplishments.

Someone asked the vice president and director of research at General Motors, Charles F. Kettering, how it happened that General Motors was making most of the diesel-electric locomotives in the country. He said, "You must have awfully good patent protection." Kettering replied: "Well, here's the reason. You see, a great many people think we're crazy. That is much better protection than any patent."

Several years ago a research engineer told me he thought I was crazy when I was teaching him my creativity technique. A week later, he became so enthused about a new gyroscope concept that he spent his evenings developing it on his own time. My technique is so simple that it is hard to believe it works so well. Basically, all I do is ask "why" and "how" over and over again, just like I did when I was young. I named these questions the "Why-How Logic Questions." I also ask several other thought-provoking questions that broaden my understanding and stimulate my creativity.

FAST AND TESTIMONIALS

After I discovered this technique, I gave a presentation on the subject in Boston, Massachusetts. During the presentation, I displayed a diagram to show the answers to the Why-How Logic Questions and demonstrated how the logic tied the answers together. I named this diagram a FAST Diagram. Most users just call my technique "FAST." I have been overwhelmed by its acceptance. Here are two comments from the hundreds who have written me:

We are confident that, if we keep to the rules of your FAST approach, the answer we seek will be found.

> —*Leon M. Turner*
> *Management Consultant*
> *John P. Young & Associates*
> *Hawthorne, Australia*

The FAST Diagram clarifies a problem and pinpoints the area to apply creativity.

> —*Richard J. Park*
> *Manager, Value Control*
> *Chrysler Corporation*
> *Detroit, Michigan*

After I discovered that my technique works on any type of problem, I stated this fact in an article. A director for the state of Pennsylvania went looking for a solution to the state's financial problems because the state was in bankruptcy or close to it at the time. After a little searching, he came upon my article, his staff applied my technique, and he wrote me the following:

> FAST reduces the time for complex analysis…One diagram may be worth more than many times one thousand words. It can be understood and appreciated by almost anyone.
>
> > —*Donald P. Goss, Director*
> > *Bureau of Systems Analysis*
> > *Commonwealth of Pennsylvania*
> > *Harrisburg, Pennsylvania*

A staff member of the Commonwealth of Pennsylvania wrote the following about the Why-How Logic Questions and FAST:

> The process of answering the formula makes "creative planning" both inescapable and so easy that program managers may not realize how creative they really are. This formula changes the brainstorming list from almost to fully complete!…An invalid answer is almost automatically apparent even to a person not skilled in the management of the program being planned…One Pennsylvania agency produced an agency plan by another method which required 638

pages. FAST would produce a more comprehensive and meaningful plan using less than 50 pages!"*

I visited the Chrysler plant in Detroit, Michigan shortly after it started to use my FAST technique. I was informed that the plant had been so successful using it that the company introduced it into its Canadian operations and was in the process of extending it to its England operations. The following was written to Sperry Univac's (a division of the Sperry Rand Corporation) division manager in Salt Lake City:

> We feel that there is no question that the results produced would have never been accomplished without the use of the FAST Diagram to stimulate and organize our thinking and to pinpoint the specific area for improvement.
>
> —*H.T. Hearon, Comptroller*
> *Chrysler Corporation*

A COMMUNICATION TOOL

An example of the effectiveness of FAST was given by Jerry Kaufman of Houston, Texas, and I quote:

> I have used FAST to improve an orthopedic procedure replacing a knee joint for the producer of prosthesis. The closing comment was that "this is the first time the medical doctors and engineers were able to communicate on the same level."

BASIC CONCEPTS

Some of the basic concepts of my FAST Creativity technique have been used throughout the world by hundreds of people who have achieved unbelievable success. As I have reviewed how these people are applying my technique, I realize that some of my original concepts *have not* been fully explained or

---

* Article supplied by John S. Hollar, Cost Reduction Program Supervisor, Commonwealth of Pennsylvania, Harrisburg.

understood; therefore, I will cover those concepts in greater detail in this book. You have the opportunity to learn them first. This book groups all my concepts together so you can learn them firsthand as you follow a variety of examples step by step.

This book has free materials available for download from the
Web Added Value™ Resource Center at www.jrosspub.com.

2

SPARKED BY FUNCTION

In 1960 I experienced a new way of thinking. I learned about this new way of thinking when I was assigned to conduct the first value engineering seminar within the Sperry Rand Corporation. Shortly after receiving this assignment, I enrolled in the first value engineering class ever taught at a university, at the University of California at Los Angeles. This was a workshop class designed to teach a technique developed by Lawrence D. Miles, a General Electric purchasing manager. The technique was called value analysis and it is applied to any type of product or service.

VALUE ANALYSIS

Value analysis is a technique that focuses "on one objective — equivalent performance for lower cost."[1] In order to achieve this objective, Mr. Miles identifies and names the functions performed by various products and services. He says that the only reason a customer purchases a product is because of the function it performs. For example, a customer purchases a lawn mower because it performs the function "cut grass." A customer purchases an electric shaver because it performs the function "remove whiskers." Every product performs or accomplishes at least one basic function.

This was an entirely new way of thinking for me. Thinking in terms of functions sparked within me new creative insight into almost everything I looked at. It opened my mind to creative opportunities beyond my experience. In the following section, I will tell you about giving names to functions. In Chapter 3, I will tell you more about my experience in discovering functions while attending UCLA.

NAMING FUNCTIONS

Names are given to functions. The first word of the name is always an active verb and the last word of the name is always a noun. "Charge battery" is the name of a function, "charge" being the active verb and "battery" being the noun. The name given to a function describes what is to be accomplished without disclosing the method of accomplishment. Figure 2.1 lists the names of four automobile functions.

A customer who is concerned about the water he or she drinks and wants to perform the function "limit impurities" may purchase a water distiller, distilled water, a filtering system, or some other type of water-conditioning system. This method of naming functions makes each name a springboard for creative development. It sparks within each individual who reads the name of a function an opportunity to visualize or imagine different ways a function can be accomplished. I know of no other tool or technique that has such a dynamic creative effect on the human mind.

Take a moment and ponder each of the names given to the functions listed in Figure 2.2. Note that the first word of each name is an active verb and the last word is a noun. As you read each name, think of what it means to you. Think of the various ways you could accomplish each function.

The reader of the function may determine how each function is to be accomplished. If I ask you to give me three ways to accomplish each of the functions listed in Figure 2.2, you probably could do it. These are actual functions named by participants who have performed valuable studies. Those who conduct value analysis studies usually take an existing product, which has been designed by someone else, and then identify all of the functions performed by that product. They list these functions either as "use functions" or "aesthetic

Start Engine
Steer Automobile
Stop Automobile
Remove CD

Figure 2.1.   Automobile Functions

Attract Attention
Collect Data
Collect Dirt
Control Deflection
Convert Energy
Create Image
Educate Students
Establish Budget
Increase Incentive
Maintain Clearance
Prevent Contamination
Protect Children
Reward Achievement
Save Time
Transmit Information
Worship God

Figure 2.2. List of Functions

functions." *Use functions* require something to be accomplished so a person will want to purchase and use the product, whereas *aesthetic functions* require something to be accomplished so one or more of the five senses will make the product more desirable than a competitive product. Examples of these two different types of functions are given in Figure 2.3.

LOOKING AT FUNCTIONS

The people who use Mr. Miles's value analysis technique are trying to maintain the product's performance and at the same time reduce its cost to the potential purchaser or buyer. Their approach is to first identify the names of all the

Use Functions	*Aesthetic* Functions
Cut Grass	Improve Appearance
Mix Ingredients	Make Convenient
Tighten Bolt	Reduce Size
Transport People	Suppress Noise

Figure 2.3. Use and Aesthetic Functions

Mount Unit
Supply Power
Blow Air
Direct Air
Remove Heat

Figure 2.4.   Partial List of Air Conditioner Functions

functions presently being performed by a given product. For example, if the product is an air conditioner, they first identify the basic function of the entire unit. They might name the basic function of an air-conditioning unit "cool environment." Then they identify all the other functions presently being performed within the entire unit; 10, 20, or perhaps 30 functions are usually identified and named by analyzing the various parts. A partial list of functions for this unit is shown in Figure 2.4.

Once all the functions have been named, all the physical parts that allow each function to be performed are identified. Since each part costs a given amount of time and money to manufacture and assemble, a cost can be assigned to each function by accumulating the cost of all the parts that collectively perform each given function. The basic function is assigned the total cost, and all other functions are assigned a portion of that total cost. Once this cost analysis task has been completed, the next step is to creatively see if other ways of performing the basic function can be conceived. The other functions that have been assigned the greatest cost are then explored to see if a more creative method of performing those functions can be conceived or discovered.

This approach divorces a person's thinking from the various parts that allow the product to exist and permits him or her to concentrate solely on functions. This method of analyzing a product usually removes 30 to 40 percent of the cost compared to 5 or 10 percent when using the normal cost reduction techniques used to analyze parts. Some of these normal cost reduction techniques are listed in Figure 2.5.

FUNCTIONS CHANGED MY LIFE

When I was taught how to name functions, it changed my life and my thinking. It changed me even more when I discovered that I could name functions for everything, even for just a few words expressed within a sentence or phrase. For example:

> Buying Parts in Larger Quantities
> Die Casting a Bracket Instead of Fabricating It
> Obtaining More Competitive Bids
> Reducing Costs by Using Cheaper Material
> Reducing Costs by Using Thinner Metal
> Using Plastic Gears Instead of Steel Gears

Figure 2.5. Normal Cost Reduction Techniques

The opposite of love is selfishness

yields the following two functions:

Express Love
Avoid Selfishness

Naming functions can do the same for you if you will take the time to learn how to name functions properly and learn the value of asking "why" and "how" for each function you name. Functions can be given names for products, procedures or processes, expressions, goals, things to do, or almost anything else, as shown in Figure 2.6.

The function name does not tell how something is to be accomplished; that is left up to the imagination and creativity of those who decide how that function is to be accomplished. However, names given to functions influence the thinking and potential creativity of those performing the analysis. When function naming is learned and done correctly, a two-word phrase offers opportunities for major development.

Because a function defines an action upon something, the verb selected must be an active verb for this relationship to exist. Active verbs suggest something

Functions	Items Associated with Functions
Cut Grass	a *product* like a lawn mower
Distribute Mail	a *procedure* within an office
Express Love	an *expression* like "I love you"
Obtain Degree	establishing a *personal* goal
Procure Book	thinking of something you need *to do*

Figure 2.6. Functions for Various Areas

Analyze
Control
Distribute
Increase
Transmit
Verify

Figure 2.7.   Active Verbs

should happen or some type of action should take place. Figure 2.7 shows some examples of active verbs. Active verbs motivate, stimulate, and energize your thinking and at the same time arouse within you your creative and inventive abilities.

SOMETHING MUST BE ACTED UPON

You need to realize that there can be no action taken unless there is something to be acted upon. A noun identified that something. You combine the active verb with a noun to form the name for the function that needs to be performed. Let's take the active verbs in Figure 2.7 and combine them with nouns to be acted upon to form the functions listed in Figure 2.8.

Keep in mind that a function describes what you want accomplished without identifying how it is to be accomplished. If you want a table cleaned, a tire fixed, or perhaps a sofa moved, those requests can be recorded as the following functions: "clean table," "fix tire," and "move sofa." These three functions do not inform you how to perform the tasks; they merely tell you what needs to be accomplished. Therefore, just by reading the name, you can use your own

Analyze Results
Control Heat
Distribute Books
Increase Income
Transmit Data
Verify Address

Figure 2.8.   Adding Nouns to Be Acted Upon

creativity to perform the tasks. Hence, a function is a creative tool in and of itself. Every time you see the word "function" in this book, think of a function as an active verb plus a noun and use it to spark your creativity.

FUNCTIONS ARE NOT LIMITED TO PRODUCTS

Even though function names initially were conceived to define work and sell aesthetic properties of products, today thousands of people are using functions to define almost everything. Consider the following statement:

> When you get up in the morning and have breakfast, you may eat ham and eggs along with some toast and drink a cup of coffee, or if you are in a hurry you may eat a doughnut along with a cold glass of milk to wash it down.

Can you state in two words what is described in this sentence? Try using an active verb and a noun to describe it. Take a couple of minutes to come up with your answer. Write it down before you read the next paragraph.

Active Verb   *Noun*

Was your active verb "eat" and your noun "breakfast" or "food" to form the function "eat breakfast" or perhaps to form the function "eat food"? These two names are functions. If you are skilled in the technique of naming functions, you might have named the function "obtain energy." Your mind becomes much more creative when you think of different ways to "obtain energy" than if you limit your thinking to "eat breakfast" or "eat food."

You will learn how to become more skilled after you have learned the basic fundamentals of this technique. There are at least 13 other functions within the above statement. Six obvious functions are listed in Figure 2.9. Seven inferred functions are shown in Figure 2.10. Figure 2.11 shows several more functions that could be named.

Notice how creative your mind becomes as you think of all the things that have to be accomplished just to eat breakfast. Your mind works the same way on anything you investigate if you think in terms of functions. You are always dealing with functions every day, even though you may not be aware of it.

Drink Coffee or Drink Milk
Eat Doughnut
Eat Eggs
Eat Ham
Eat Toast

Figure 2.9. Obvious Functions

Cook Eggs
Cook Ham
Crack Eggs
Make Coffee
Pour Coffee
Pour Milk
Toast Bread

Figure 2.10. Inferred Functions

Energize Toaster
Insert Bread
Spread Butter
Supply Cup
Supply Glass
Supply Plate
Supply Utensils

Figure 2.11. Additional Functions

Functions are things you think about before you actually decide how you are going to do them.

DECISIONS: A FORM OF CREATIVITY

A function as simple as "eat eggs" requires one or more other functions to be performed. This function depends upon the decisions you have to make when

Egg Omelet
Poached Eggs
Scrambled Eggs

Figure 2.12.   Partial List of Choices

you decide to eat eggs. You have to decide from among a number of choices, such as those listed in Figure 2.12.

SUPPORTING FUNCTIONS

Obviously, there are many more choices that could be made, such as having an egg casserole or egg salad or possibly going out to eat, in which case you still have to inform the server of your choice regarding eggs. Let's say you decide to have a poached egg. This decision produces the function "poach egg." This function is a supporting function of "eat eggs." In other words, if you weren't going to eat an egg, you wouldn't have to poach an egg. This new function creates a branch of its own functions in a logic diagram. This will be discussed in greater detail in Chapter 8. It is sufficient at this time to state that *any function that requires you to make a decision rather than express your creativity produces a function that supports the function you are analyzing.* Therefore, you can think of decision functions as supporting functions.

When a function is properly written that does not require a decision, it always stimulates creativity. If it fails to do this, you know it is not a good function! Every good function is capable of stimulating at least one creative way of executing or performing that particular function. Since you have the opportunity to decide how you want to accomplish a function each time you name one, you are continually opening the door so you can express your creativity. The writing of functions requires skill, in-depth thinking, and experience. Mr. Miles stated:

> Intense concentration, even what appears to be overconcentration of mental work on these functions, forms the basis for unexpected steps of advancement of value in the product or service.[2]

The verb and noun you select to define a function have a direct bearing upon your ability to expand your mind into new areas; therefore, it is very important

Obvious Functions	Mind-Expanding Functions
Eat Breakfast	Obtain Nourishment
File Letter	Store Documents
Purchase Apples	Purchase Fruit or Purchase Food
Say Hello	Display Courtesy

Figure 2.13. Obvious and Mind-Expanding Functions

that you devote considerable thought in naming each function. Figure 2.13 lists several obvious functions along with corresponding mind-expanding functions.

NOTES

1. Miles, Lawrence D., *Techniques of Value Analysis and Engineering*, 2nd edition, McGraw-Hill, New York, 1972, p. 3.
2. Ibid., p. 26.

3

DISCOVERING FUNCTIONS

Now that you know a little bit about functions, let me tell you more about the value engineering workshop I attended at the University of California at Los Angeles. After we were taught several steps that are required to effectively conduct a value engineering study using value analysis, the main step known as naming functions was presented. This step taught us that every component of a mechanical assembly was there for a reason and that our assignment was to find out what each component did and express that fact using an active verb and a noun. Our instructor for this technique was Ed Heller. He told us that each verb-noun we recorded was the name we were to give to a function. He also told us that if we could name the functions for the mechanical assemblies we were assigned, we could probably improve our assigned product or project and at the same time reduce its manufacturing cost by as much as 30 percent or more. The improvements were to be accomplished by inserting each function name into the How Else Question developed by Lawrence Miles and then creatively answering each question. The How Else Question reads as follows:

How else can *this function* be accomplished?

I had never seen or heard of this method of thinking before, and the thinking about functions fascinated my mentality. As five of my teammates and I worked on our mechanical assembly, we soon named a number of functions and began thinking of other ways of performing those same functions. Within two or three days, we had all kinds of ideas for how our project could be improved. It was amazing how creative we became as we inserted those functions into this question.

We were able to eliminate some parts, and we redesigned other parts so they were capable of performing several functions.

DISCOVERING BASIC FUNCTIONS

After the class was completed, I continued to ponder this function concept. One thing in particular seemed to be lacking within the steps we were taught. There was no clear-cut method of identifying the basic function of the entire project. It appeared that most value engineers just picked what they thought was the basic function, which in most cases probably turned out to be correct. That wasn't good enough for me. This deficiency continued to dominate my thinking for the next two or three years as I taught this newfound function technique to seminar participants.

In the meantime, I continued to be amazed by how productive I was as I helped participants analyze the functions they named for their projects. I was always asking questions about their functions. Finally, one day I was able to formulate a question that isolated the basic function from a list of functions of an existing complex assembly. Instead of guessing which function was the basic function, I selected the function I thought was the basic function and inserted that function into this new formulated question. The question reads as follows:

> If I didn't have to perform *this function*, would I still have to perform any of the other functions listed?

If the answer was no for all the other functions listed, then I knew I had indeed identified the basic function. Not only had I identified the basic function, but the function I had identified caused all of the other functions to come into existence. Even though the list of functions in Figure 3.1 is an obvious example,

Blow Air
Cool Environment
Direct Air
Mount Unit
Remove Heat
Supply Power

Figure 3.1.   Partial List of Functions for Air-Conditioning Unit

it demonstrates what I have stated. For example, if I am analyzing an air-conditioning unit and I select the function "cool environment" as the function I insert into this question, it becomes the basic function if I do not have to perform any of the other functions listed.

Also, *the method selected to perform the function "cool environment" causes the other functions listed to come into existence.* With the basic function identified, I could then insert it into the How Else Question and search for other creative solutions. If a cool mountain stream could be utilized to cool the environment within a room, the list of functions would be completely different. I needed a name for this technique, so I called it the "Basic Function Determination Technique."

As I applied the logic of this question, I seemed to have a hidden talent for knowing what other questions to ask and how to evaluate the answers. I had many opportunities to do this because I was still conducting seminars at the time for the Sperry Rand Corporation. We generally had six mechanical or electrical assemblies for each seminar that pertained to a missile system we were building for the United States Army. I had ample opportunities to try all of my logic questions on a variety of different projects using participants to answer the questions for me. These people came from various backgrounds, had many technical skills, and had a lot of personal experiences and knowledge.

DISCOVERING HOW I WAS THINKING

Right after we had completed a very successful seminar, my supervisor, Corwin S. Grey, said to me: "Your creativity works for you, but it won't work for anyone else because they don't know what you are doing or how you think. Why don't you try putting down on paper how you think?" He said this because I had helped several teams obtain unusual results by asking them a number of questions about the functions they had listed. When I tried to record how I was thinking, I was amazed. Most of the time, I was just asking, "Why do we have to perform this function?" and "How is this function accomplished or proposed to be accomplished?" I thought to myself, "Doesn't everyone ask why and how when they analyze a problem?"

ANSWERS EXPRESSED AS FUNCTIONS

The next thing I realized was that there was a relationship that existed between the answers and the Why-How Questions I asked and the function I was analyzing. The result of this observation *permitted me to express my answers to the Why and How Questions about functions also as functions.* Not only that, but I could logically tie these functions together.

WHY-HOW LOGIC

Up to this point in time, value engineers just made a list of functions, and from the list they developed, they selected what they thought was the basic function. There was no attempt to show any relationships between functions. Also, *function analysis wasn't even known as a discipline at that time.* Whenever I selected any given function in my list of functions that was not the basic function and then answered the question

Why do I want to perform ***this function***?

I frequently discovered that my answer was already expressed as a function in my list of functions. And if it was not on the list, I could express my answer as a function. When I did this, I realized I was actually moving one level higher in my logic thinking because of what I had learned when I coined my Basic Function Determination Question. Many times, a new higher level function also became the new basic function. This Why Question also taught me that my answer to this question caused the function I inserted to come into existence. I needed a way of visually showing this relationship, so I placed this new function within a rectangle at the left of the function I inserted, as shown in Figure 3.2.

This seemed to me to be the most logical place to post the answer since the answer, expressed as a function, was more important and actually caused the inserted function to come into existence. Therefore, when I looked at the visual relationships I had created by showing functions on a flat surface before me, I could see the higher level functions first, since by habit I normally read from left to right. This was desirable because the highest level function always possesses the greatest potential for creative development. It also causes every

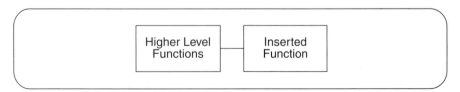

Figure 3.2. Visually Displaying Higher Level Functions

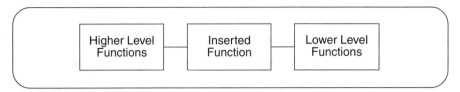

Figure 3.3.   Visually Displaying Lower Level Functions

function to the right of it to come into existence. The next thing I discovered was that when I asked the question

How is ***this function*** actually performed?

this answer could also be expressed as a single function or as several functions. These new functions were entirely dependent upon the function I inserted into the How Question. In other words, if I did not have to perform the function inserted in this How Question, any new functions produced by answering this question would not exist. Therefore, these new functions were considered to be lower level functions or dependent functions. I placed these new functions to the right of the inserted function, as shown in Figure 3.3. The higher level functions as well as the lower level functions will be discussed in greater detail later.

After this discovery, I started to ask "why" and "how" of these new lower level functions. Each time I repeated this process, I generated additional functions and my understanding of my project increased. This was exhilarating to me because I began to explore areas and information that I had not even considered up to that point. I needed a name for this method of analyzing functions, so I called it the Function Analysis System Technique, which is better known by its acronym, FAST.[1]

CONCEPT OF FAST DIAGRAMMING

The result of writing down the functions as they relate to each other generated a visual diagram which showed *how each function is performed* by merely observing the functions posted immediately to the right of any given function. By the same token, if one desired to know *why a given function is required,* the function posted at its immediate left provided the answer. It soon became evident that the Why-How Logic relationships only existed and worked if what was being analyzed or investigated was recorded using the verb-noun method of naming functions. When I realized this, I decided, *I had indeed discovered something that was very unique.* Then I thought, "Who would believe something so simple could be of any value since people have been asking 'why' and 'how' since the beginning of time?"

Since my technique appeared to stimulate a lot of creativity whenever I asked the Why-How Questions, I decided to write an article about my technique in as much as I was recording how I did my thinking for my boss. My seminar participants and staff members had demonstrated how effective my thinking process had been as we recommended tremendous changes in the products we analyzed. These changes had the potential to reduce the cost of some assemblies we were purchasing and manufacturing by more than 30 percent. The production cost of one unit we worked on later was reduced by 80 percent. I needed a sample project to demonstrate these Why-How Logic relationships, so I selected a household item that everyone uses, an incandescent lightbulb. This sample project will help you understand the steps of this technique presented in Chapter 8. It is a project that anyone can comprehend.

In order to show the relationships between the various functions used to describe this lightbulb, I created a diagram similar to Figure 3.4. This diagram is easy to read and understand. As I formulated the diagram to show these relationships, I needed a name for it and decided to call it a *FAST Diagram.* FAST, as mentioned earlier, is the acronym for Function Analysis System Technique.

The diagram in Figure 3.4 allows the functions of the lightbulb to be displayed in one drawing. This arrangement permits lines to be drawn between different functions, thereby showing how each function is related to other functions to the left or right of the function being considered. If you follow the line to the left of a function, it connects to the function that causes that function to "come into being." If you follow the line to the right, it connects to those functions that describe the methods by which that function is accomplished in the present design or a proposed design.

Every function in a FAST Diagram should possess these same relationships if the relationships have been verified by asking the Why-How Logic Questions.

FAST TESTIMONIALS

A few more comments from those who have used FAST follow. For the first three quotes, I am only able to provide the location where they were originated.

> The rigorous logic of FAST turned value analysis from a game of playing with words into a legitimate, organized problem-solving system.
>
> —*Dayton, Ohio*

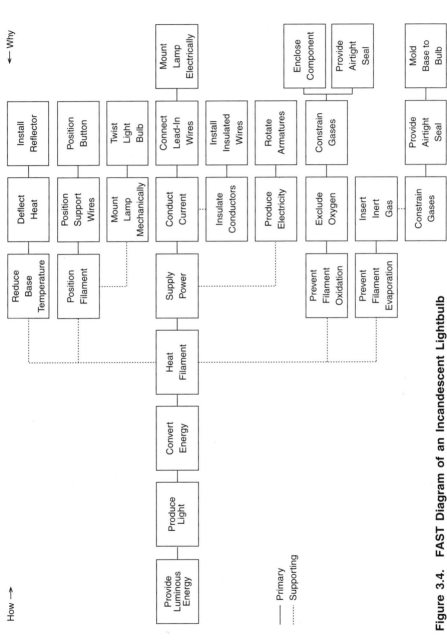

Figure 3.4. FAST Diagram of an Incandescent Lightbulb

Those who have been trained in FAST...regularly report that they first understood their problem as they began to develop their own FAST Diagram...The process by which a FAST Diagram is prepared is a nearly ideal communication link.

—Rochester, New York

The use of FAST Diagramming as the tool for creating the understanding needed to properly plan for the future is a fact which we cannot overlook and is an important problem definition tool which can be applied to any problem requiring a creative solution.

—Glasgow, Kentucky

We have undertaken plant layout studies, labor utilization, complete manufacture and packaging of panty hose, procurement systems, inventory control systems, corporate cash flow,...invoicing systems, debtors' systems, credit payment systems....When confronted with some of the types of projects mentioned, we are confident that, if we keep to the rules of your FAST approach, the answer we seek will be found.[2]

—Leon M. Turner
Management Consultant
John P. Young & Associates

Because a FAST Diagram is a logic diagram, its very nature causes it to come into its own when used to analyze complex problems...We have applied the FAST Diagram technique to over 100 projects. We are also using the FAST approach in our application task forces to develop new designs, improve old products, analyze operational procedures, and invent new approaches. Results are extremely satisfactory. The FAST Diagram...clarifies a problem and pinpoints the area to apply creativity. It then helps to sell the idea by providing a concrete plan to demonstrate and guide discussion.

—Richard J. Park
Manager, Value Control
Chrysler Corporation

To my knowledge, FAST is the only available technique for real product structural analysis.

—C.A. Lopez Padilla
Manager, Manufacturing Engineering
General Electric Argentina

For understanding and dealing with functions we, today, have the most remarkable system of its kind devised by man. It is the Function Analysis Systems Technique, originated by Charles Bytheway, now being understood, supported and used by hundreds and extended by a few.

—Lawrence D. Miles
Inventor of Function Terminology
Founder of Value Analysis

FAST is a technique for enhancing more productive thinking and creativity and at the same time helps solve any type of problem. It may also be employed to analyze and solve problems of a more private nature, problems like how to strengthen the family, how to purchase a new car, how to live within one's income, or how to raise good children. Hundreds and probably thousands of various types of projects could be listed. Chapter 5 lists a number of completed projects. Five different projects are demonstrated in the chapters that follow.

Most of the examples presented herein describe, explain, and demonstrate how a particular step is to be performed. In addition, an attempt has been made to show the value of performing each step.

Every project using this technique can be expanded into a larger project just by asking "why" of the higher level functions and asking "how" of its lower level functions. Since this is true of any project, you must be realistic or practical whenever you use this technique. By that I mean you have to put limits on how much you are going to expand any given diagram. By the same token, most productive creative opportunities exist and are exposed to your understanding only when you extend the diagram to the left by asking "why" of higher level functions. Therefore, always extend the diagram to the left until you name a function that sparks your interest. If you obtain one, dedicate your energy and creativity to that function and its lower level functions. The following chapters give examples of how this is accomplished, and a couple of the examples demonstrate when just a branch of a developed diagram should be selected for creative development.

As you develop a FAST Diagram, you need to realize that each time you ask "how" of a function, your stimulated creativity is required to describe either how something is performed that already exists or how else you can perform that same function. Each time you do this, you are moving to lower creative opportunity levels on the diagram. When you ask "why" of a function, you are increasing your creative opportunities because your answers cause you to formulate higher level functions. These higher level functions bring all lower level functions into existence. However, the lower level functions only identify one method of performing the new higher level function. An alternate method of performing a higher level function often is simpler, easier to implement, and costs less and may require a new diagram to be constructed.

THE INVENTOR OF FUNCTIONS

The initial concept of functions as defined herein was invented by Lawrence D. Miles as World War II conflict was coming to an end, while he was working as a purchasing agent for the General Electric Company. Since materials required for the war effort were in short supply for companies that were supplying goods for the general public's consumption during those war years, many of General Electric's products had to be discontinued until a substitute material could be found. Mr. Miles began to correspond with potential suppliers by asking if they had something that General Electric could use in place of rubber, steel, and other materials that were allocated strictly for military use during those World War II years.

Sometime later, Mr. Miles started to describe to his potential suppliers what he needed by using functions expressed by an active verb and a noun. This was just one of the steps he used to obtain new suppliers that could supply different materials and products to General Electric. Not only was the cost of some of these products reduced, but it also made it possible for General Electric to supply them once again during those war years. He discovered that the most productive technique was when his suppliers offered him products or parts that performed the same function but at tremendously lower cost. This approach was so beneficial to General Electric that Mr. Miles was promoted to manager and trained other people within the company so they could do the same thing. His verb-noun concept of naming functions, along with his other concepts for reducing costs, was included in the training program he developed. He named his technique *value analysis*. His technique was adopted later by the United States Navy and named *value engineering*.

FUNCTIONS AND FAST

Hundreds if not thousands of people have named functions for the projects they have worked on using Mr. Miles's technique and have organized and creatively analyzed those functions using my FAST diagramming technique. *Analysts are now able to understand their projects better, collect and fill in missing information, communicate with other project participants better, arrive at creative solutions in much less time, and display a diagram that allows decision makers to understand and implement their proposals quicker. These two techniques,*

Aim Jet	Generate Sales	Pump Blood
Allocate Resources	Guarantee Reliability	Quantify Group
Analyze Results	Identify Variances	Quote Job
Announce Events	Improve Communication	Recommend Solution
Appoint Specialists	Increase Convenience	Reduce Friction
Approve Art	Increase Reliability	Reduce Inventory
Assign Responsibility	Influence Customer	Regulate Flow
Assign Tutor	Inform Operator	Release Product
Assure Convenience	Inject Insulin	Remove Temptation
Audit Books	Input Code	Replace Alcohol
Avoid Confusion	Inspect Product	Replenish Inventories
Become Obedient	Install Unit	Requisition Materials
Communicate Data	Instruct Personnel	Retrofit Unit
Compress Gas	Interview Client	Salvage Material
Conduct Current	Invoice Customer	Scope Program
Contact Customer	Justify Investments	Seek Help
Convert Signal	Lease Equipment	Select Clients
Cool Fluid	Load Equipment	Select Material
Correct Deficiencies	Make Report	Select Movie
Create Confidence	Monitor Patient	Select Seating
Create Design	Motivate Salespeople	Select Table
Create Drawings	Negotiate Changes	Shield Radiation
Credit Account	Order Food	Simulate Environment
Describe Process	Overcome Addiction	Store Data
Destroy Cells	Paint Parts	Store Glycogen
Determine Potential	Perform Tests	Supply Adrenaline
Determine Quantity	Please User	Supply Fluid
Develop Plan	Position Cutter	Support Weight
Display Instructions	Predict Reliability	Test System
Elevate People	Prepare Proposal	Track Performance
Entice User	Prevent Embarrassment	Train Salespeople
Entrain Air	Prevent Flow	Transmit Information
Explore Options	Process Changes	Update Document
Filter Noise	Propel Vehicle	Utilize Staff
Finalize Agreement	Protect Operator	Win Contract
Forecast Costs		

Figure 3.5.   List of Functions from Actual Projects

naming functions and FAST diagramming, make this all possible. The names of some of the functions displayed in these diagrams are listed in Figure 3.5. Many more could be added.

The reader should realize that the thinking that takes place in naming a function is vital to a creative experience and that functions should not be named by looking down a list generated by someone else. As you look over the list of functions in Figure 3.5, you will read function names that pertain to a large variety of projects which range from management to software, hardware, communications, medical, etc. Sometimes the same function must be recorded in two or more segments of a given project or the function must be more specific to give clarity. In these situations, a middle name or modifier is added.

FUNCTIONS MADE SPECIFIC

Figure 3.6 lists functions that may require modifiers. There is no set rule, but the fewer the modifiers, the greater the opportunity to stimulate thinking and creativity because modifiers generally tend to restrict or narrow the scope of a person's imagination, unless you have the ability to role-play.

Functions	Functions Made More Specific
Aim Jet	Aim Power Jet
Aim Jet	Aim Exhaust Jet
Confirm Schedule	Confirm Order Schedule
Construct System	Construct Electrical System
Construct System	Construct Mechanical System
Create Drawings	Create Detail Drawings
Create Drawings	Create Assembly Drawings
Destroy Cells	Destroy Cancer Cells
Determine Potential	Determine Patent Potential
Establish Practices	Establish Quoting Practices
Estimate Costs	Estimate Developmental Costs
Issue Certificate	Issue Evaluation Certificate
Manufacture Items	Manufacture Customized Items
Prevent Embarrassment	Prevent Student Embarrassment
Process Changes	Process Emergency Changes
Publish Forecast	Publish Economic Forecast
Receive Instructions	Receive Disposition Instructions
Requisition Products	Requisition Standard Products
Resolve Problems	Resolve Delivery Problems
Select Components	Select Electrical Components
Sensitize Entrainment	Sensitize Air Entrainment
Train Personnel	Train Customer Personnel
Train Salespeople	Train New Salespeople

Figure 3.6. Functions Made More Specific

PROJECTS FOR THE READER

Consider the following six projects to see if you understand the concept of functions. Take a few minutes to generate all of the functions you can think of, and then compare your list to the functions listed in Chapter 4 for these same projects. Remember to use active verbs combined with nouns.

1. Driving an automobile
2. Using a word-processing program
3. Making a cake
4. Riding a bicycle
5. Changing disc brakes on an automobile
6. Preventing crime in your neighborhood

This exercise will give you experience in thinking in terms of functions. The best way to learn how to name functions is to actually name some functions yourself. Approach each project by visualizing in your mind all the things you would have to do if you were going to teach someone else how to perform the tasks required by each project. You may not be able to complete all six projects, but you should try to do at least two or three. Keep in mind that the functions you write down do not have to be recorded in a step-by-step sequential order. You should write down any function that comes to your mind. Record as many as you can.

Richard J. Park lists 280 verbs and 363 nouns that were taken from actual projects conducted by him and his associates. These are listed in his book, *Value Engineering: A Plan for Invention.*[3] His list pertains to value engineering–type projects.

One day while I was looking for a part to perform a certain function at an electrical supply shop, I noticed a part in one of the bins and thought to myself, "That's a great idea. Why didn't I think of that?" As I inspected the part a little more closely, I discovered that it was not what I thought it was. Just like that, I had an idea for a new invention. I came home and made five different models which utilized this new concept. The more you think in terms of functions, the more you look at common things from a different point of view.

DO YOUR OWN ANALYSIS

Some of my FAST technique's basic concepts have been used by hundreds of people throughout the world for almost 40 years. Now, for the first time, all of the concepts and the development of this new method of thinking are presented in this book so you can quickly and easily do your own analysis and have fun doing it. It is exciting because you never know where it will lead you.

NOTES

1. Bytheway, C.W., "Basic Function Determination Technique," SAVE Proceedings, Fifth National Conference, Vol. II, 1965, pp. 21–23; also Bytheway, C.W., "FAST Diagrams for Creative Function Analysis," SAVE Communications and *Journal of Value Engineering*, Vol. 71-3, March 1971, pp. 6–10.
2. Bytheway, C.W., "Innovation to FAST," SAVE Proceedings 1972, North Central Regional Conference, 1972, p. 6.1.
3. Park, Richard J., *Value Engineering: A Plan for Invention*, St. Lucia Press, New York, 1999, pp. 317–321.

4

WHY-HOW LOGIC

At the end of Chapter 3 it was suggested that you get some experience in writing functions. Six projects were presented for you to name all the functions you could think of. I trust that you have tried your skill on at least two or three of them. The entire FAST Creativity technique deals with functions from beginning to end. Therefore, it is vital that you learn how to name functions.

NAMING FUNCTIONS

It is assumed that the exercise at the end of Chapter 3 helped you become accustomed to naming functions and thinking in terms of functions. See how well you did by comparing your list to Figures 4.1 through 4.6.

Activate Lights	Observe Speed Limit
Check Gas Gauge	Observe Speedometer
Check Oil Gauge	Prevent Accidents
Check Tire Pressure	Push Brake Pedal
Close Door	Release Emergency Brake
Control Gas Pedal	Set Emergency Brake
Fasten Seat Belt	Shift Gears
Learn Driving Rules	Start Engine
Observe Rear View	Turn Off Lights
Observe Safety Rules	Turn Steering Wheel

Figure 4.1. Project One: Driving an Automobile

Block Text	Load Program
Bold Text	Move Text
Capitalize Letters	Number Pages
Center Text	Print Document
Change Columns	Relocate Paragraph
Change Fonts	Replace Words
Change Font Size	Retrieve Deletes
Change Margins	Retrieve Document
Check Spelling	Retrieve Text
Create Columns	Right Align Text
Create Paragraph	Save Document
Delete Text	Start Computer
Delete Words	Switch Documents
Exit Program	Turn Columns Off
Find Synonyms	Turn Columns On
Find Words	Underline Text
Indent Paragraph	View Two Documents
Left Align Text	

Figure 4.2.   Project Two: Using a Word-Processing Program

Add Decoration	Purchase Ingredients
Apply Frosting	Read Directions
Clean Working Area	Select Baking Pan
Coat Pan Surface	Select Mixing Bowl
Follow Directions	Select Recipe
Heat Oven	Set Temperature
Make Frosting	Spread Frosting
Measure Ingredients	Test Cooked Cake
Mix Ingredients	Use Spatula
Prevent Crumbling	Wash Utensils
Prevent Sticking	

Figure 4.3.   Project Three: Making a Cake

Obviously, many more functions could be listed for every project. If we took the time, we could organize the functions of the last project (Figure 4.6) into a very effective program to prevent crime in neighborhoods; however, to do so without the participation and cooperation of people within a neighborhood probably would end up just a program on paper. If the neighbors, community

Adjust Seat   Keep Balance
Change Gears   Lubricate Parts
Check Tires   Minimize Effort
Create Torque   Obtain Exercise
Develop Muscles   Stop Bicycle
Guide Bicycle   Support Weight
Inflate Tires   Transport People

Figure 4.4. Project Four: Riding a Bicycle

Bleed Brakes   Lower Automobile
Check Fluid Level   Observe Pad Position
Compress Plunger   Pour Break Fluid
Dry-Lubricate Pads   Remove Bolts
Elevate Automobile   Remove Caliper
Install Caliper   Remove Disc Pads
Install Disc Pads   Remove Wheel
Install Wheel

Figure 4.5. Project Five: Changing Disc Brakes on an Automobile

Convict Violators   Record Auto Plates
Detect Lewd Acts   Record Child Abuse
Enlist Neighbors   Record Drug Sales
Hold Watch Meetings   Record Drug Use
Investigate Noises   Record Graffiti
Know Neighbors   Record Misconduct
Know Officers   Record Prostitution
Know Their Relatives   Record Spouse Abuse
Observe Clothes   Record Trespassing
Observe Features   Remove Graffiti
Observe Strangers   Report Collected Info
Organize Parents   Report Poor Lighting
Petition Officials   Turn Porch Lights On
Record Alcohol Use

Figure 4.6. Project Six: Preventing Crime in Your Neighborhood

leaders, officers, and others were to use these few functions and worked together to develop a FAST Diagram, they could end up with a program that everyone could understand, and everyone would know why each type of crime exists and how they could help prevent those crimes. In Chapter 1, I talked about children asking "why" and "how" about almost everything. I am now going to talk about these two questions and show how they are related to each other.

WHY-HOW LOGIC

The Why and How Questions were asked concerning functions in Chapter 2. Also, the answers to those two questions were converted into functions. Each time this occurred, an attempt was then made to ask the same questions of any new functions which were formed from the answers. When repeated, this process allowed a diagram to be constructed, which I named a FAST Diagram. The diagram is really a logic diagram. The logic used is known as the *Why-How Logic*. This logic allows all the functions in a diagram to be tied together into a cause and effect relationship.

Everyone usually asks *why* and *how* as they engage in any thinking exercise, but no one has ever before attempted to establish a relationship between the answers to these two questions. It was not until I tried to write down how I was thinking that I realized I had actually established a relationship between the answers to these two questions. As I did so, I was able to understand the project I was working on a little better. Not only that, but I could also determine the correctness of the results when I asked "why" of a how function and "how" of a why function by the answers those questions produced. It also became obvious when information was missing.

CONVERT ANSWERS TO FUNCTIONS

I will now show you how you can establish these relationships for yourself as you analyze any subject you wish to investigate. The *key* to doing this is to *convert your answers into functions*. The names you give these answers tell why you are trying to do something and describe how you are presently accomplishing something or proposing to accomplish it. These answers are logically related to each other.

When you describe how a given function is performed, you bring into existence new functions. When you ask why a given function exists, you identify the function that caused that particular function to come into existence.

These relationships always exist between functions. Your job, when you use my FAST Creativity technique, is to name the functions and make sure this relationship exists between the functions you name. If you perform this task well, I promise you that many creative doors will open for you. *The logic that ties two functions together is known as Why-How Logic.* If the logic holds, then your thinking is considered to be correct.

START WITH A SIMPLE FUNCTION

Let's consider something every individual looks at or thinks about perhaps several times every day: their teeth or other people's teeth. In order for this logic to work on a subject involving teeth, we need to name a function that has something to do with teeth. We'll start with the function "clean teeth." Then we'll ask the Why Logic Question of this function, which reads as follows:

Why must ***clean teeth*** be performed?

When we answer this question, we may respond "Teeth must be cleaned so a person will be attractive." Reducing this answer to a function, we can give it the name "obtain attractiveness." Since every time the Why Logic Question is asked we are actually moving to a higher level function, we will use the procedure we developed in Chapter 2 to demonstrate this. In other words, we will place this new function to the left of the "clean teeth" function, as shown in Figure 4.7.

VERIFYING YOUR ANSWER

To verify if "obtain attractiveness" is the correct answer, all we have to do is insert this new function into the following How Logic Question:

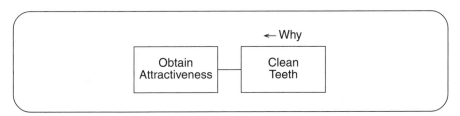

Figure 4.7.   Asking "Why" of Function "Clean Teeth"

How is *obtain attractiveness* actually accomplished?

The answer to this question must yield the function "clean teeth." If it does not, the logic does not hold. Do a person's clean teeth make a person attractive? If they did, unattractive people would be cleaning their teeth all the time. Obviously, the answer is no! Therefore, the logic does not hold. The reason "obtain attractiveness" is not correct is because it is more than one level above the level where the logic holds. This suggests that information is missing and more analysis is required.

From a creative standpoint, however, "obtain attractiveness" opens many more doors than "clean teeth" because it is two or three logic levels higher. At this point, we have two options when we are applying the Why-How Logic. The first option is to insert the new function, "obtain attractiveness," into the How Logic Question and expand it until one of the lower branches of the diagram includes the function "clean teeth." This option requires more skill and creativity and is the approach that is generally used once one is skilled in the technique. We will use that approach in some of the example projects in the following chapters. However, in order to demonstrate the value of the Why-How Logic, we will concern ourselves with the other option.

The other option is to repeat the first Why Logic Question on the same function until we obtain an answer where the logic does hold. Therefore, let's answer the same question once again:

Why must *clean teeth* be performed?

A more obvious answer for this question is the function "prevent cavities." Now we will check to see if this is the correct answer. When we insert this function into the How Logic Question, we obtain the following question:

How is *prevent cavities* actually accomplished?

Obviously, "clean teeth" answers this question correctly. So just by answering the How Logic Question when "prevent cavities" has been inserted, we say the logic has been verified in both directions. We could have answered the How Logic Question with the function "remove bacteria." If this had been the case, we would have had to do a little more in-depth thinking in order for the logic to hold in both directions. This subject is considered in greater detail in Chapter 8. For now, we will stick with the two functions "prevent cavities" and "clean teeth." The function "prevent cavities" which we inserted into the above How Question does more than just verify that "clean teeth" is correct logically; it also suggests we think of all the other things we can do to prevent cavities.

USING YOUR CREATIVITY

Rarely does just one function answer the How Logic Question completely. Therefore, we can become creative at this point by repeating the How Question and try to think of different answers for this question or we can insert the function "prevent cavities" into the how else form of the How Question as follows:

How else can *prevent cavities* actually be accomplished?

When we answer this question, other ways of preventing cavities come to mind, such as "avoid eating sweets," "eat wholesome foods," "schedule regular checkups," "floss teeth," "use fluoride toothpaste," and "rinse food particles." So just by answering the How Logic Question we are able to generate six and possibly more new functions and at the same time verify that "clean teeth" agrees logically.

As the number of functions increases, generally the function names are recorded on small rectangular slips of card stock. This permits us to show their relationships with respect to each other by where they are physically placed on a flat surface. The higher level functions are placed at the left and the lower level functions at the right. The arrangement of these new functions is shown in the FAST Diagram in Figure 4.8.

When we look at this FAST Diagram and want to know how to prevent cavities, we look in the direction of the "how" arrow and find the answer. Likewise, when we want to find out why a given function is required, we just look in the direction of the "why" arrow for the answer. The logic diagram is really a question diagram and an answer diagram.

USING WHY LOGIC

Now that we have displayed the information from this first How Question exercise, we need to check to make sure the logic holds for the six new functions in the "why" direction. If we insert the first two and the last three of these new functions we just created into the Why Logic Question, we obtain the question:

Why must *avoid eating sweets, eat wholesome foods, floss teeth, use fluoride toothpaste, and rinse food particles* be performed?

The answer for all five functions is the same: to "prevent cavities." Since this function is located at the immediate left of these five new functions, the

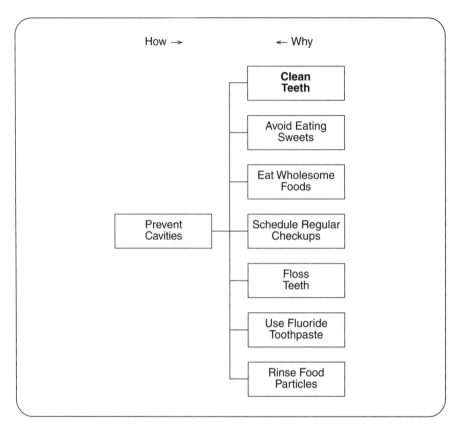

Figure 4.8. FAST Diagram for "Prevent Cavities"

logic is said to hold in both directions. However, when we insert the third new function, "schedule regular checkups," into this Why Logic Question, it reads as follows:

Why must *schedule regular checkups* be performed?

This question could be answered by saying that by scheduling regular checkups we "prevent cavities," but "scheduling regular checkups" will never "prevent cavities." It doesn't matter how many you schedule. Therefore, the logic between these two functions does not hold. This example demonstrates why many people fail to obtain the full benefit of the Why-How Logic. You must check every function to see if the logic holds in both directions. By that I mean you must check the How Logic by asking the How Logic Question of a function at the left and determine if all functions at its immediate right actually tell how that function is accomplished or performed. Then check each one of those functions at the right to see if the function immediately at its left tells why it needs to be performed.

VERIFICATION QUESTION

If you have trouble deciding whether or not the logic holds in the why direction, then insert the how function you are investigating along with the why function you are trying to verify into the following Verification Question:

Does **this how function** help **its why function**?

If we insert the appropriate functions from the above discussion, we obtain:

Does **schedule regular checkups** help **prevent cavities**?

If the answer is no, then the logic does not hold. If the answer is yes, then it does hold. Obviously, we still get the same "no" answer, so the logic does not hold. Those who understand and rigorously apply the Why-How Logic have new creative doors open for them because they move into areas they have not considered before.

TESTIMONIALS OF FUNCTION LINKS

Here are three more testimonials that refer to the Why-How Logic as the formula:

> The process of answering the formula makes "creative planning" both inescapable and so easy…This formula changes the brainstorming list from almost to fully complete!…An invalid answer is almost automatically apparent.*

> The importance of the FAST approach is that it graphically displays function dependencies and creates a process to study function links while exploring options to develop improved systems.
>
> *—Jerry Kaufman, President*
> *J. Kaufman Associates, Inc.*
> *Houston, Texas*

---

* Article supplied by John S. Hollar, Cost Reduction Program Supervisor, Commonwealth of Pennsylvania, Harrisburg.

The great virtue of Bytheway's breakthrough is that it preserved the simplicity and concentrated on relevance of the verb and noun definition while tackling the thorny problem of cause and effect...I urge that we keep FAST diagramming simple, straightforward, and robust.

—*Carlos Fallon**

As stated in Chapter 2, every time we name a new function, we need to ask the Why-How Logic Questions of it. When we do this, our diagram begins to grow as new creative doors open and our understanding of our subject broadens. Also, the number of functions grows exponentially. Consider the five new functions we ended up with after discarding "schedule regular checkups." If we ask the How Logic Question of these, we could develop five new branches to our diagram. Obviously, the diagram can become extremely large if every branch is expanded.

LOGIC DIAGRAMS AND FAST TREES

In Chapter 2, a FAST Diagram for an incandescent lightbulb was demonstrated, and a similar diagram for "prevent cavities" has been started in this chapter. These diagrams were created by placing higher level functions at the left and lower level functions at the right and tying them together with appropriate straight lines. We need a method to visually display the functions and the logic that tie all the functions together for large projects.

When lots of functions are generated, the FAST Diagram method of creating a logic diagram becomes too cumbersome and too difficult to keep track of our thinking and where each function should be placed unless we have a large flat surface or wall on which to post the functions. The computer permits us to do the same thing rather quickly and at the same time retain a record of where we have been.

Therefore, another method of creating a logic diagram will be discussed, which I introduced when I presented my initial paper in Boston in 1965. I called it a Functional Family Tree, but it will be called a FAST Tree in this book. It is also created by asking the Why-How Logic Questions, but the functions are typed into a word-processing file. Each line is used to record only one function. It is a good idea to set your tab spacing for 0.3 inches if you have a large project.

Start your project by tabbing in from the left margin about four tab spaces and enter your first function. If you have 20 functions, you will need 20 lines to record your logic diagram. Instead of placing new why functions directly at

---

* *Value Analysis,* 2nd revised edition, Triangle Press, Irvine, Texas, 1980, pp. 199 and 203.

```
                    Prevent Cavities
                    Clean Teeth          ↖
         How        Eat Wholesome Foods    Why
             ↘      Avoid Eating Sweets
                    Use Fluoride Toothpaste
                    Rinse Food Particles
                    Floss Teeth
```

Figure 4.9.   FAST Tree for "Prevent Cavities"

the left, they are typed one tab space to the left and above the function entered into the Why Logic Question. Likewise, new how functions are typed one tab space to the right and below the function entered into the How Logic Question. Since new how functions can be more than one function, they are all listed at the same tab position below the why function that caused them to come into being. Because this may be a little confusing, a FAST Tree that contains the same information as the FAST Diagram shown in Figure 4.8 will be presented, except it does not include the function "schedule regular checkups" since that function did not comply with the logic. The FAST Tree for this project at this stage of development is shown in Figure 4.9.

This computerized system allows the display to show how the logic ties the functions together and permits the diagram to expand vertically instead of horizontally. This logic diagram method of displaying functions is called a FAST Tree because it grows vertically like a tree and utilizes the same concepts as the FAST Diagram, which grows horizontally. It provides a number of other advantages over the FAST Diagram, which will be identified later.

We will not investigate the function "schedule regular checkups" further, since it does not comply with the Why-How Logic. However, often it is beneficial to set a function like this aside and apply the Why-How Logic to it later in an effort to find out where it actually fits. For now, we will concentrate our discussion on those functions which demonstrate the application of the Why-How Logic.

LOOKING FOR CREATIVE OPPORTUNITIES

The main objective in any analysis is to move into areas one has not yet considered. At this point, we do not know what or where those areas are, but our approach is to expand our logic diagram upward and to the left in search of a higher level function that will motivate us and excite our creativity. There-

Create Perfect Teeth
Prevent Cavities ↖
How **Clean Teeth** Why
↘ Eat Wholesome Foods
Avoid Eating Sweets
Use Fluoride Toothpaste
Rinse Food Particles
Floss Teeth

Figure 4.10. Initial FAST Tree for "Create Perfect Teeth"

fore, we will continue to ask the Why Logic Question of the higher level functions.

The next question for this sample problem is as follows:

Why must ***prevent cavities*** be performed?

The ideal answer to this question is "create perfect teeth." We would like to add this new function to our already developed logic diagram. Once again, the computer provides a means of accomplishing this, and at the same time the computer allows us to retain a record of what we have thought about up to this point. All we have to do is scan up in our document and locate the logic diagram we developed earlier and copy it. In this case, it is the FAST Tree shown in Figure 4.9. Then we scan down to the end of our document or file and paste a copy of that logic diagram. Next we add a line above our highest entry and type in "create perfect teeth" at the correct tab stop, as shown in Figure 4.10. Then, of course, we have to change the figure number to indicate that it is different from the previous one.

APPLYING WHY-HOW LOGIC TO NEW FUNCTIONS

Next we check to make sure the new function, "create perfect teeth," agrees in the How Logic direction by asking the following question:

How is ***create perfect teeth*** actually accomplished?

One way to "create perfect teeth" is to "prevent cavities," but does "prevent cavities" "create perfect teeth"? No! Does that mean the logic does not hold?

At first glance, it appears that it does not. We have to realize that the How Logic Question is a creative question and suggests that there probably exists more than one function that has to be performed to accomplish a given function completely.

Obviously, many more functions could be listed in the FAST Tree in Figure 4.10 under "prevent cavities," such as "remove bacteria" or "kill bacteria." However, the six functions listed will be sufficient for demonstration purposes.

SURE TEST FOR HOW FUNCTIONS

Therefore, if we are not sure a particular how function agrees with the logic, as in the above example, we should ask the Verification Question:

Does *this how function* help *its why function*?

For our example, this question becomes:

Does *prevent cavities* help *create perfect teeth*?

The answer is yes; therefore, it is verified in the why direction. We can only use the Verification Question to verify if a function is logically correct when we already have an answer in the why direction. The Why Logic Question is normally used when no entry has been posted in the why direction. You should always remember that the why direction for a FAST Tree is one tab space to the left and up, and the how direction is one tab space to the right and down.

Once we have verified the logic in the why direction, we need to reinsert the answer to the Why Logic Question into the How Else Question so we can utilize our creative abilities once again. This creative exercise allows us to add a few new functions to our logic diagram, as shown in Figure 4.11, by asking the following question:

How else can *create perfect teeth* actually be accomplished?

If we were blessed with a set of irregular teeth, then the first thing we might think of is the function "straighten teeth." If, on the other hand, we are missing two or three teeth, we might think of "install implants," "purchase bridge," or "purchase dentures." Any one of these could make us look like we had perfect teeth. We are more interested in how our teeth look than how they got that way. Here we have made the assumption that we are not striving for perfect teeth,

Create Perfect Teeth
Prevent Cavities ↖
How **Clean Teeth** Why
↘ Eat Wholesome Foods
Avoid Eating Sweets
Use Fluoride Toothpaste
Rinse Food Particles
Floss Teeth
Straighten Teeth
Install Implants
Install Crowns
Purchase Bridge
Purchase Dentures

Figure 4.11. Partial FAST Tree 1 for "Create Perfect Teeth"

but we want to look as if we have perfect teeth. If a tooth has been broken or chipped, we might consider fixing it with a crown.

VERIFICATION QUESTION

Since the function "create perfect teeth," which brought these five new functions into existence, is already posted in our logic diagram shown in Figure 4.11, we can bypass the Why Logic Question and insert these new functions into the Verification Question:

> Do *straighten teeth, install implants, install crowns, purchase bridge,* and *purchase dentures* help *create perfect teeth*?

Since the answer is yes, we can now move on and ask the How Logic Question of these new functions. We will limit our analysis to "straighten teeth" for this particular example. When we do this, we obtain the new function "receive orthodontic treatment."

The computer allows us to quickly copy and insert this new function where it belongs logically. We are still able to retain the original because all we do is copy the original, paste it below, and then modify it by inserting the new function and change the figure number, as shown in Figure 4.12. Note that the "why" and "how" notations and corresponding arrows are not indicated since

Create Perfect Teeth
   Prevent Cavities
      Clean Teeth
         Eat Wholesome Foods
         Avoid Eating Sweets
         Use Fluoride Toothpaste
         Rinse Food Particles
         Floss Teeth
   Straighten Teeth
      Receive Orthodontic Treatment
   Install Implants
   Install Crowns
   Purchase Bridge
   Purchase Dentures

Figure 4.12. Partial FAST Tree 2 for "Create Perfect Teeth"

by this time you should know the logic arrangement of functions when constructing a FAST Tree. The Verification Question can then be asked of this new function, which indicates that the logic is correct.

SEARCHING FOR CREATIVE OPPORTUNITIES

If, perchance, you are an orthodontist, you could ask the How Logic Question of the function "straighten teeth" and expand that branch of the logic diagram. As you explore these areas, you might name a function that sparks your creativity and end up with an entirely new method of straightening teeth. You might discover a way to grow teeth where they have been extracted by using stem cell research or DNA research.

Let's expand the logic by asking one more question of the highest level function:

Why must create perfect teeth be performed?

The answer I like is "obtain attractiveness." You may recall that we obtained this function when we first started this discussion. At that time, it was pointed out that "obtain attractiveness" was several levels above "clean teeth." Let's verify if it is a good function first, and then, if it is correct, we will post it three

Obtain Attractiveness
    Create Perfect Teeth
        Prevent Cavities
            Clean Teeth
            Eat Wholesome Foods
            Avoid Eating Sweets
            Use Fluoride Toothpaste
            Rinse Food Particles
            Floss Teeth
        Straighten Teeth
            Receive Orthodontic Treatment
        Install Implants
        Install Crowns
        Purchase Bridge
        Purchase Dentures

Figure 4.13.   Initial FAST Tree for "Obtain Attractiveness"

levels above "clean teeth." The capital letter "Q" will be used to designate a question, and the capital letter "A" will be used to designate its corresponding answer throughout this book.

> Q   Does *create perfect teeth* help *obtain attractiveness*?
> A   Yes

When this new function is added to our logic diagram in Figure 4.13, it automatically opens a number of creative doors. Those creative doors become more evident when we insert "obtain attractiveness" into the How Else Question, which then becomes:

> How else can *obtain attractiveness* actually be accomplished?

The results of answering this creative question are shown in Figure 4.14.

FAST EXPANDS OUR THINKING QUICKLY

Note that in order to expand the FAST Tree, all we had to do was ask the How Else Logic Question of some of these higher level functions. Also note that we just started this entire analysis by writing down the function "clean teeth."

Obtain Attractiveness
  Create Perfect Teeth
    Prevent Cavities
      Clean Teeth
      Eat Wholesome Foods
      Avoid Eating Sweets
      Use Fluoride Toothpaste
      Rinse Food Particles
      Floss Teeth
    Straighten Teeth
      Receive Orthodontic Treatment
    Install Implants
    Install Crowns
    Purchase Bridge
    Purchase Dentures
  Improve Figure
    Obtain Breast Implants
    Lose Weight
  Improve Eyesight
    Receive Corrective Eye Surgery
  Change Hair
    Receive Hair Implant
    Color Hair
    Receive Hair Styling
  Change Facial Features
    Change Nose Structure
    Change Jaw Structure
    Remove Wrinkles

Figure 4.14. Partial FAST Tree 1 for "Obtain Attractiveness"

Consider how long it would take to obtain this same information if we were to just ponder the subject of cleaning our teeth.

We have by no means fully investigated this subject yet, but this demonstrates what can be accomplished in a short period of time merely by asking the Why-How Logic Questions and the How Else Logic Question. This basically is what FAST is all about.

Now that we have developed a FAST Tree for "obtain attractiveness," it is easier to understand and follow the logic if we convert it into the FAST Diagram shown in Figure 4.15. Either method of displaying the information is considered acceptable. Obviously, the FAST Tree is much easier to create and to change;

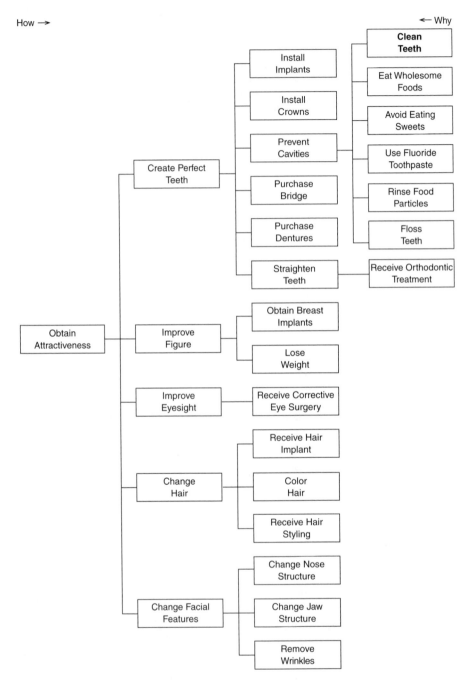

Figure 4.15. Partial FAST Diagram for "Obtain Attractiveness"

however, the FAST Diagram is easier to comprehend because all you have to do is follow the lines that connect the functions.

Many more functions could be added to either of these two logic diagrams merely by asking the Why-How Logic Questions of any function. Note that if you select any function in one of these diagrams, you can find out "why" that function is necessary by merely following its logic construction. Likewise, its construction allows you to discover "how" each function is to be accomplished.

For example, if you select "change facial features," you can see that "obtain attractiveness" identifies why this function is necessary. If you want to know how "change facial features" is accomplished, you discover that there are three ways listed, namely "change nose structure," "change jaw structure," and "remove wrinkles."

You should always verify if all of the functions posted in your logic diagram are correct logic-wise by asking each appropriate Verification Question, such as:

Does ***change nose structure*** help ***change facial features***?

If you feel you are not as attractive as you would like to be, you could expand this logic diagram by adding additional functions. This little exercise has been designed merely to teach you how to apply the Why-How Logic and to teach you how to verify that your answers are correct.

SUMMARY

Within a few minutes we were able to generate a lot of information just by initially asking the Why Logic Question of the function "clean teeth." As we did this, we were able to find the function that caused "clean teeth" to come into existence. That function was "prevent cavities." Just by identifying this new function we were able to consider many different ways of accomplishing the task of preventing cavities by asking the How Else Logic Question and utilizing our creativity. The effect of doing this resulted in a list of new functions.

LOGIC KEEPS ANALYSIS FOCUSED

The logic diagram demonstrates that a *cause and effect* relationship exists between two functions and permits your thinking to be concentrated and centered on your subject. Since every new function possesses these same cause and effect relationships with other functions, you do not have to wait for additional inspiration to expand your subject matter. The logic diagram does it for you. All you have to do is continue to ask the Why-How Logic Questions of any new functions.

Those who have applied the logic can verify that this relationship keeps them focused on the subject they are analyzing and that it saves them time and money. This technique gathers essential information about the subject and allows it to be read and understood by almost anyone, as the following testimonials confirm:

> FAST reduces the time for complex audit and analysis of management planning to perhaps one-third of the time it would otherwise take, and at the same time increases the quality of the planning process and plans presentation.
>
> —*Donald P. Goss, Director*
> *Bureau of Systems Analysis*
> *Commonwealth of Pennsylvania*

> One Pennsylvania agency produced an agency plan by another method which required 638 pages. FAST would produce a more comprehensive and meaningful plan using less than 50 pages!*

So many people would not make claims like these unless they have actually used the technique and found out for themselves that it stimulated their creativity and permitted them to be more productive.

UNIQUENESS OF LOGIC DIAGRAMS

Every entry in a logic diagram is used to form two questions; in other words, each entry can be used to ask why this function is required or how this function is accomplished. Also, every entry in a logic diagram answers the How Question for the function at its left and also answers the Why Question for the function or functions at its right. *Can you think of any other diagram capable of forming two questions and answering two other questions with only one entry of two or three words?* There is none that I know of. Think about it —

---

* Article supplied by John S. Hollar, Cost Reduction Program Supervisor, Commonwealth of Pennsylvania, Harrisburg.

a logic diagram that contains 40 entries has the capability of asking 80 questions as well as answering 80 questions. The same information would take pages and pages to document, and considerable dialogue would be required to explain the relationships. Also, it would be extremely difficult to keep track of those relationships. Once you understand the construction of a FAST Diagram or a FAST Tree, those relationships can be read and understood by anyone.

If you merely record the obvious and fail to get the Why-How Logic to agree in both directions, you are just going through the motions and will not experience the full benefit that can be realized from this technique. In other words, the technique will be of little value to you.

Let me warn the reader that this thinking process requires energy, persistence, and in-depth thinking; however, the rewards obtained far outweigh the effort required to perform the task. I had no idea that "clean teeth" would lead into investigating other ways of making a person more attractive or considering the possibility of using stem cell research and/or DNA to grow new teeth. These last two concepts may seem ridiculous at this time, but they may spark an idea and could possibly motivate a researcher to consider them.

A COMMUNICATION TOOL

Most conflicts exist during an analysis because of poor communication between the parties involved when several people are working together on a given project. The main benefit of using the Why-How Logic is the thinking and communication that take place as the participants try to reason out the logic and arrive at meaningful answers. This activity requires participants to communicate with each other until they agree on the wording and meaning of each name they give a function.

Whenever one person's answer to a Why Question is different from the answer given by another person, a dialogue always follows. Eventually an agreement in semantics is achieved or it is recognized that their perceptions of the problem are so varied that they must ask some of the other logic questions before they can hope to come to a unified understanding of that particular segment of the subject.

As they continue the analysis process, the lines of communication are opened so that each person involved can understand the other person's point of view. After a particular study performed by doctors and engineers who were trying to improve an orthopedic procedure to replace a knee joint, the concluding remarks included the following statement:

This is the first time the medical doctors and engineers were able to communicate on the same level.

—Jerry Kaufman, President
J. Kaufman Associates, Inc.
Houston, Texas

As each person responds to the logic questions concerning each function, a wide range of answers emerges, resulting in a discussion dialogue wherein each party is able to understand and know why the other individuals involved think the way they do. Many times, people want the same thing, but the words they use to present their ideas mean something completely different to other people because of their experience in life.

Each member of every task force team has different skills, education, talents, experiences, associates, companions, children, schoolmates, friends, neighbors, relatives, political and religious leanings, cultural traditions and customs, etc., and their individual thinking will take them in many different directions. What has happened to people throughout their entire lifetime will influence their response to the FAST logic questions.

The Why-How Logic Questions, when properly applied, usually disclose these differences. Without these questions, it is similar to a religious teacher telling a class of young children that "God drove Adam and Eve out of the Garden of Eden." The children are probably visualizing God in the front seat of an automobile and Adam and Eve in the back seat. Obviously, the teacher is not communicating with the children. The same thing happens with adults when they discuss functions. This dilemma can be resolved rather quickly by asking each person to tell the other participants how he or she thinks each function is to be performed and why he or she thinks each function is necessary.

As each person does this, new functions may come to light because of individual points of view. Information that is missing immediately becomes evident. It also opens new avenues for thinking that would not otherwise be considered or explored. Even when verification of the logic is being discussed, differences have to be resolved. Most people agree that this whole exchange of what each person is thinking as he or she answers these questions is a powerful communication tool.

This book contains a stimulating and systematic creative and innovative thinking approach which synergistically utilizes team members' ingenuity and originality to quickly broaden their comprehension and effectiveness in communicating with each other while solving simple and complex problems. This approach permits members of a team to utilize their logical and intuitive analyzing capabilities as they reason with fellow team members to arrive at meaningful functions which satisfy the logic constraints associated with the FAST thinking process.

MISSED OPPORTUNITIES

When you fail to verify if your answers are correct, you are missing opportunities to expand your understanding and also missing chances to stimulate your creativity. When the logic does not agree, it should suggest to you that a mental discussion should take place within you and/or among several other people you are working with until the logic holds. As each discussion takes place, other opinions should be expressed freely until a consensus is reached.

LOGIC DIAGRAMS: FAST TREES AND FAST DIAGRAMS

The logic diagram developed to demonstrate the Why-How Logic in this chapter utilized a word-processing program on a computer. The logic diagram used in this discussion grows vertically and is called a FAST Tree. The logic diagram in Chapter 2 expanded the functions horizontally and is called a FAST Diagram. They are basically the same thing and contain the same information. Just follow the logic and you will know how to read either one of them. Both methods of displaying these relationships were introduced in my original presentation in Boston, Massachusetts in 1965. The only difference then was the fact that a computer filled an entire room and cost thousands of dollars; now almost everyone can afford a computer, so a FAST Tree is the best way to apply my FAST technique. Remember that you don't need a computer to develop a FAST Tree; all you need is a pencil and a piece of paper.

Both methods of presenting logic diagrams are utilized in the chapters that follow. The FAST Tree has many advantages over the FAST Diagram, especially if you are working alone or if you have several people helping you who have their own computers and you are communicating with each other over the Internet. Three of those advantages are as follows:

- You have the opportunity to have several other people record their answers to the Why-How Logic Questions at the same time on different computer terminals. This allows each person to do his or her own thinking instead of being influenced by what someone else expresses.
- It allows all the information generated during the extensive thinking process to be kept in sequential order. It is not necessary to look through

a pile of paperwork to find a thought or an idea that sparked your creativity. You can enter right in your computer any thoughts you may have as they come to you. Since the computer file grows as your analysis progresses, these ideas and thoughts that have been recorded are easy to locate, which allows you to remember what sparked them.

■   A corporation or government agency can now develop programs, invent new military or civilian products, and solve problems over the Internet by applying the FAST Creativity procedural steps. This saves travel and other associated expenses and valuable time lost while traveling to and from destinations.

The FAST Diagram has some advantages when working with a team of participants. When the FAST Diagram is being used, functions are generally recorded on small rectangular cards, as was done at the beginning of this chapter. The cards allow the functions to be appropriately moved and positioned on a flat surface as the discussion takes place. The FAST Diagram is also easier to see by a large group of participants gathered in the same room. After the functions and logic have been verified, lines are drawn to connect the answers to the Why and How Questions regarding a particular function. Also, segments of the diagram can be displayed easily and assigned to different groups for further study and in-depth analysis. The final completed diagram may then be used to obtain approval and funding for the concept expressed by the functions. *Even though FAST Diagrams have been the norm in the past, it is recommended that a person duplicate the teams' work on a computer in the form of a FAST Tree. The results can then be projected so everyone can see as trees are copied, pasted, and then changed as participants agree on changes.*

5

SELECTING
A PROJECT

A variety of projects have been analyzed successfully using the questions and logic of FAST. It doesn't seem to matter what type of project you decide to select. FAST allows you to gather a lot of information within a very short amount of time. The project can be software, hardware, or anything else you elect to investigate or analyze. It can be used to develop new designs, improve old products, analyze processes or procedures of any kind, invent new approaches, clarify a problem and pinpoint an area to apply creativity, and even can be used to teach or just learn about a subject. You can even prepare a talk and use it to make sure you have adequately covered the subject. It can be used to establish training procedures or any other type of procedure or to analyze something that already exists or a concept that just exists within your mind. The following two statements demonstrate the variety of projects that have been successfully selected by two companies:

> We have undertaken plant layout studies, labor utilization, complete manufacture and packaging of panty hose, procurement systems, inventory control systems, corporate cash flow,...invoicing systems, debtors' systems, credit payment systems.

> —*Leon M. Turner*
> *Management Consultant*
> *John P. Young & Associates*
> *Hawthorne, Australia*

We have applied the FAST diagram technique to over 100 projects. We are also using the FAST approach in our application task forces to develop new designs, improve old products, analyze operational procedures, and invent new approaches. Results are extremely satisfactory.

—Richard J. Park
Manager, Value Control
Chrysler Corporation
Detroit, Michigan

Figure 5.7 at the end of this chapter lists many more projects that have been analyzed successfully, many of which are nonhardware oriented, such as energy conservation, an elementary school system, a marketing and sales operation, operations within a title insurance company, or why people steal cars.

You may already have a project in mind to analyze, to investigate, or to study. If you do, you still should select one of the methods listed below that best fits that project. However, if you do not have any project in mind, read through the various methods, select one, and see how easy it is to get started.

METHODS FOR SELECTING PROJECTS

- **Method 1** — Write down and describe any item, subject, object, part, assembly, product, procedure, or process you wish to analyze.
- **Method 2** — Write one, two, or three sentences about the project. The information contained within these sentences is used to start the project.
- **Method 3** — Whenever a team is assigned to analyze a product or a piece of equipment which is normally sold to a customer, client, or consumer, answer the six questions for Method 3 listed in Figure 5.1. The information contained within the answers to these questions defines your project.
- **Method 4** — When you don't have a clue as to what to select as a project, answer one or more of the 15 questions for Method 4 listed in Figure 5.2. Then select all answers that pertain to the same area of concern and use the information contained therein for your project. You may end up with three or more projects if you answer all of the questions.
- **Method 5** — Answer the four questions for Method 5 listed in Figure 5.3 for projects that have problems that need correcting. Then, from the information contained within those answers, formulate one or more projects.

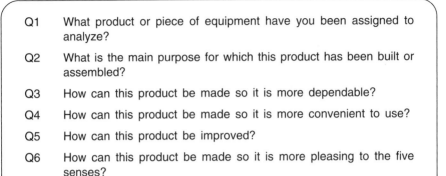

Figure 5.1.   Questions for Method 3

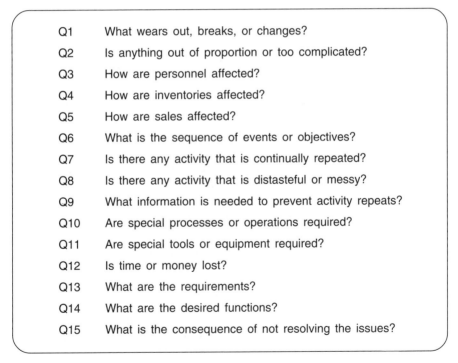

Figure 5.2.   Questions for Method 4

> Q1   What problem shall we discuss?
>
> Q2   Why do you think this is a problem?
>
> Q3   Why do you think a solution is needed?
>
> Q4   What is there about this problem area that disturbs you?

Figure 5.3. Questions for Method 5

The five methods of selecting a project should make it easy to select a project that will open the door to your creativity. Keep in mind that no two projects should be treated the same. Several projects have been selected to demonstrate the various steps of this creative technique in an effort to illustrate that each project may lead you in a variety of different paths. It is suggested that you study all projects so as to get a feel for the flexibility this technique extends to you. Most projects are best performed by a team of four or five people.

USING METHOD 1

Sometimes the project selected involves a product or article you would like to understand better, such as a lightbulb. When an article or an assembly is selected, you just write down the name of each part or the items associated with it and then assign the project the name of the item or assembly. Even though a lightbulb is a small assembly, it is selected for demonstration purposes because the parts used to make a lightbulb perform many functions. It also can be used to demonstrate how to construct a logic diagram. As construction takes place, the logic involved helps increase understanding of the design criteria used to manufacture this product. Primary path functions and supporting functions are defined as higher and lower level functions are discussed.

Project 1: Lightbulb[1]

The lightbulb shown in Figure 5.4 consists of a base rim, some insulation, a glass bulb, a stem, a stem press, lead-in wires, support wires, a filament, a heat-deflecting disc, and a center contact. Any product can be selected as a project, such as an entire automobile, its transmission, its suspension system, or an entire manufacturing plant. If larger projects are selected, then only the major units or assemblies would be listed when defining the project initially.

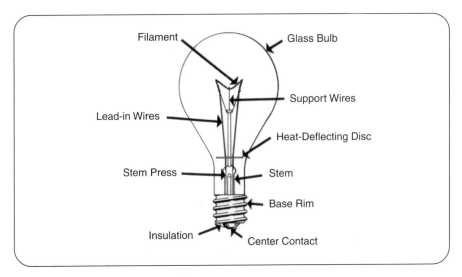

Figure 5.4. Lightbulb Assembly

Project 2: Timing Device[2]

All of the components for this timing device are mounted within the housing shown in Figure 5.5 and exploded in Figure 5.6. This design was based on a World War II timing mechanism. When the FAST technique was applied to this project, the analysis was limited to the parts mounted within the housing. The clock assembly or escapement mechanism was supplied by another manufacturing firm.

The primary parts consisted of an arming wire, arming pin, O-ring, arming lever, start/stop plunger, timing disc, clock, timing lever, timing lever shaft, sear lever, cocking shaft, and the firing pin. The functions performed by these parts contribute directly to the method selected to perform the basic function of the entire unit. These functions are called the primary path functions. All other parts assist these parts in performing their functions. This project demonstrates what can be accomplished if emphasis is placed on the basic function and all other functions within the primary path.

There are usually secondary functions that exist in any project, each one of which has its own logic path. For example, if the function "release firing pin" is required to perform the basic function for the method selected, then this function is automatically in the primary logic path. By the same token, the firing pin cannot be released unless it is first cocked so it can be released, which indicates that a supporting function also must exist. This will be explained in greater detail when this project is demonstrated in Chapter 9.

Figure 5.5. Timing Device Assembly

There are many different kinds of hardware-type projects. Some projects have many parts and others consist of only one part.

USING METHOD 2

Method 2 only requires a sentence or two to provide the information to start a project. From these few facts, a volume of information can be generated as this creativity technique is applied. The following project starts by just writing down two sentences.

Project 3: Love

The opposite of love is selfishness. The opposite of love is trying to control another person.

—*Arthur E. Millican*

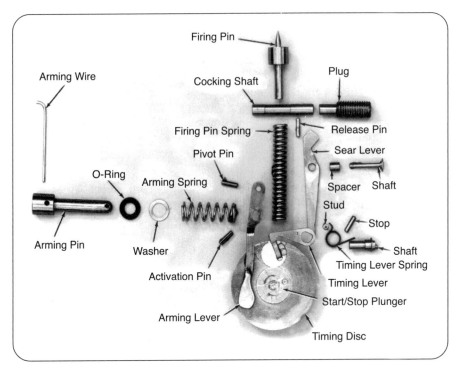

Firing Pin
Arming Wire
Cocking Shaft
Plug
Firing Pin Spring
Release Pin
Pivot Pin
Sear Lever
O-Ring
Arming Spring
Spacer   Shaft
Arming Pin
Stud
Stop
Washer
Shaft
Activation Pin
Timing Lever Spring
Timing Lever
Start/Stop Plunger
Arming Lever
Timing Disc

Figure 5.6.   Parts within the Timing Device

USING METHOD 3

Method 3 is for hardware-type projects which are purchased by consumers or special clients and are usually supplied through distributors. The hardware or equipment normally is mass produced and must be reliable and competitive in quality, price, and attractiveness. Such hardware projects also are usually analyzed by four or five team members using the FAST concepts. Chapter 6 discusses how to select the proper team members for projects where dependability, convenience, enhancing the product, and pleasing the senses of the customer are concerned.

Project 4: Three-Ton Heat Pump

This project is a three-ton heat pump. The team members assigned to this project were from engineering, manufacturing engineering, finance, purchasing, and marketing. The five questions listed for Method 3 and their corresponding answers form the nucleus for starting this project, as shown below:

Q1   What product or piece of equipment have you been assigned to analyze?

A1   We have been assigned to analyze a company's three-ton heat pump, which has the capability to heat, cool, and dehumidify a living space.

Q2   What is the main purpose for which this product has been built or assembled?

A2   Make people comfortable.

Q3   How can this product be made so it is more dependable?

A3   Make sure it doesn't deteriorate or have to be repaired very often.

Q4   How can this product be made so it is more convenient to use?

A4   Make the mounting simple and the controls readily accessible.

Q5   How can this product be improved?

A5   Make it more efficient.

Q6   How can this product be made so it is more pleasing to the five senses?

A6   Make it more attractive and make it so a normal-size person can operate it.

USING METHOD 4

No projects have been selected for demonstration purposes using Method 4, but the nucleus for this method is the same as Method 3. The answers to the 15 questions for Method 3 define the project. Keep in mind that all answers that pertain to the same area of concern form a project. Several projects usually can be obtained when all 15 questions are answered. Always use this method when you are unable to identify a project by considering the other four methods.

USING METHOD 5

Sometimes you may decide to consider a project that has several problems for which a solution is desired. You might have several projects from which to choose. This method is particularly effective when you have four or five other people willing to assist you in analyzing your project. When these conditions exist, make sure you answer all four questions for Method 5 as completely as possible. Then use the answers as the nucleus for your team's project.

Project 5: Military Communication Device

The project selected is a communication device used by the military to display, analyze, and transmit field conditions. It is a portable electronics package easily carried, inserted, and removed from a variety of host vehicles. The main components are a structural case containing the system's electronics, a removable hard drive, and a backup battery. Also included in the product are a display, keyboard, and interface cables.

Q1 What problem shall we discuss?
A1 The current communication system has acquisition and sustainment costs that exceed projected annual acquisition plus ownership allocations.

Q2 Why do you think this is a problem?
A2 Field failures have exceeded normal operating costs.
■ Water intrusion in the power unit reduces operational life.
■ Rough handling causes bezel button failure.
■ The interface between components degrades reliability.
■ Vibration causes electrical connectors to disengage.
■ System lockups and reboots occur too frequently.

Q3 Why do you think a solution is needed?
A3 Continued high cost will affect replacement quantities.

Q4 What is there about this problem area that disturbs you?
A4 Failure to correct the problems could lead to unacceptable field performance.

The information in these four answers is more than enough to initiate this project.

The FAST technique will work on anything you wish to investigate. A wide variety of projects which have been successfully completed were referred to in the testimonials in Chapters 3 and 4. Figure 5.7 is only a partial list of these and other successfully completed projects.

Advanced Data System	Marketing and Sales Operation
Air-Conditioner Compressor	New Product Design
Alcohol Recovery Program	New Product Education Program
Blind Rehabilitation Center	Operator Station
Career Planning	Product Assurance Study
Casting Processes	Product Development Process
Chapter Development Activities	Production Control Study
Coin Box Assembly	Program Planning
Conference Planning	Purchasing Systems
Copy Machine	Self-Propelled Lawnmower
Customer Order Processing	Service/Sale Study
Drill Carrier	Shipping Department
Dynamometer Operations	Student Paper Contest
Elementary School System	System Configuration Center
Energy Conservation	Thermocouple Study
Factory Rebuild Study	Title Insurance Operations
Flare Timer	Truck Assembly Process
Flux Valve	Value Control
Grenade Launcher	Vehicle Serviceability Study
How to Prevent Auto Theft	Why People Buy Cars
Hydraulic System	Why People Steal Cars
Keyboard Project	Youth Assistance Organization

Figure 5.7. Completed Projects

NOTES

1. Bytheway, C.W., "Basic Function Determination Technique," SAVE Proceedings, Fifth National Conference, Vol. II, 1965, pp. 21–23.
2. Bytheway, C.W., "Simplifying Complex Mechanisms During Research and Development," SAVE Proceedings 1968 International Conference, 1968, pp. 233–242.

6

PARTICIPANTS

Generally, you want to consider working with other people to assist you when you use FAST. How often have you been so close to the subject you were analyzing that you couldn't see or consider what other people were seeing and considering? No matter how many times you ask yourself the same questions, you get the same answers. Read the following message and count the number of Fs:

> FINISHED FILES ARE THE RESULT OF YEARS OF SCIENTIFIC STUDY COMBINED WITH THE EXPERIENCE OF MANY YEARS.

How many Fs did you count? I went over this message several times and could only count three. When I was informed that there are six, I couldn't believe it. I searched and searched and was sure that was wrong. Then I was informed that I was not considering the word "of" as I did my search. Sometimes the same type of thing happens when asking the Why-How Logic Questions; therefore, it is a good idea to have other people assist you whenever you use the FAST technique.

ASK OTHER PEOPLE TO PARTICIPATE

Every time you talk about or try to explain a certain point to someone else, you immediately realize that you understand that point just a little better. If you ask other people the Why-How Logic Questions about functions, you immediately think just a little deeper, which allows you to evaluate their answers with greater interest and understanding. This activity forces you to be more attentive and causes you to see those "Fs" you missed when you reread your own answers.

Insert a given function into one of the logic questions and ask someone to answer that question for you. The people you approach probably will not realize that you are asking them a question about a function. If you have to, inform them you are trying to teach them how to think FAST or how to create a FAST Diagram. Their curiosity probably will get them to comply with your request. For example, you could supply the following five functions: reflect energy, heat element, transfer energy, protect operator, and cook food. Then ask the people you approach to tell you "how" they would do each one of these five functions. Also ask them to describe "why" they would want to do each one of these five functions if they were required to. They will probably wonder what you are up to, but don't tell them. Just tell them to give you their best answers to these ten questions.

Another excellent way to get other people to aid you in your creative endeavor is to supply them with a list of functions similar to those developed for making a cake or driving an automobile in Chapters 3 and 4, along with the Why-How Logic Questions. Tell them to insert each of the functions into these two questions where *this function* appears and give you the best answers they can think of. You never know what point of view people will take when they study these questions, especially when you do not disclose what project you are working on. They invariably will disclose avenues you would not otherwise consider. Try it. You'll like the results.

You can make these same requests by corresponding over the Internet. People's answers may surprise you because their experiences throughout their entire lives will be used to respond to these questions. A chat room on the Internet also may be used by several individuals who are miles apart or perhaps on different continents. Consider asking these same people to become part of your task force team after you teach them the basic principles of this technique.

The discussing of different points of view and the asking of questions work the same way for every member of your task force team. When you ask the Why-How Logic Questions concerning a given function, people's individual responses will take them in many different directions. We will discuss this in a later chapter in considerable detail.

When several individuals do their own analysis and develop their own logic diagrams, the functions are always just a little different in one way or another, even though they are working on the same project. Some people name functions that fail the Verification Question test without even realizing it. Functions that do not agree logically are still good functions but require additional discussion with other participants until they can be properly placed in the logic diagram.

Without the interchange of ideas, these functions probably never would be considered and opportunities for creativity would be missed.

Also, the words people use to name functions have different meanings to different participants. This fact helps participants to enter into worthwhile discussions whenever two or more people work together. A function forces the focus to be on the subject at hand and is always centered on a particular portion of the project. If the wording is completely different or even if it is exactly the same, the answers to the Why-How Logic Questions may be different, which always results in a discussion until a consensus is reached. These activities are always beneficial. For example, if the function "transfer energy" is being discussed, one person may think of electricity and another person may think of transferring energy to the pedals of a bicycle.

FOUR DIFFERENT LOGIC DIAGRAMS

Logic diagrams are developed by following a step-by-step procedure. There are 13 steps to be considered whenever a logic diagram is being constructed. These steps are outlined sequentially in Chapters 8 and 13. Chapter 15 gives a brief summary of the entire procedure, along with a Procedure FAST Tree.

Step 2 of the procedure deals with the selection of participants. This section of this chapter deals with the interaction of the team leader with other people as the diagram is being constructed, including how this interaction should take place, since it is not covered in much detail in other chapters. When the analysis is to be performed by only one individual, the diagrams are called *Individual Logic Diagrams,* and the individual performing the analysis automatically becomes the team leader. There are three ways the team leader is involved with other people. Two of these engage participants who are capable of corresponding with each other and the team leader over the Internet.

The first of these computer-related methods involving other people is known as creating *Step-by-Step Diagrams.* In other words, the team leader selects the participants and instructs them how they are to proceed. Each individual starts with Step 3 (identify initial functions) of the procedure and completes that step before sharing and discussing his or her results with the other participants. As soon as the results have been exchanged, discussed, and merged and agreement has been reached, the next step is performed in the same manner. The combined result of all participants is always used to start the next step. This is repeated over and over again until all 13 steps have been completed and combined.

The second computer-related method involving other people is an individual effort performed by each participant using the Individual Logic Diagram method. However, each participant must perform Step 3 and share his or her results with the team leader, who in turn shares those results with the other members of the team. The entire team must agree on the initial functions before completing Steps 4 through 11 as an individual effort. At that time, the team members' completed logic diagrams in the form of FAST Trees are transmitted to the team leader, who in turn transmits all copies to the other participants. Each participant then merges the logic diagrams together into a single diagram. During this merging process, each participant is free to communicate with all other participants. When all participants have completed this task, their results once again are transmitted to the team leader, who in turn distributes them once again. These diagrams are known as *Composite Merged Diagrams.* A final merging then takes place, followed by a final combined effort to perform procedural Steps 12 and 13 as a team.

The third method involving other people is known as the *Normal Logic Diagram* method, which usually creates a FAST Diagram. It is usually constructed by four or five people working together and assembled in the same room. Sometimes a project is divided into subprojects and a separate group is assigned to analyze one segment of the total project. It is recommended that a FAST Tree be typed into a word-processing program that duplicates the activity of the team and the results projected so all may see and participate freely.

INDIVIDUAL LOGIC DIAGRAMS

You may decide to develop several logic diagrams by yourself before you ask other people to participate with you. Perhaps the best way to accomplish this is to select a small object you have around your house, such as a pen, a pencil, a lock and key assembly, a can opener, or perhaps a stapler. You might try just writing down a sentence and see how many functions you can identify within its content and then develop a logic diagram from those functions. It is recommended that you read the rest of this book before you start such a project so you have a better idea of how to approach different types of projects. Try having a friend do the same project you select and then compare your results.

The individual analysis method is easy to do. Just follow the procedure outlined in Chapters 8 and 13. Once you have formulated names for two or three functions, you can move on to the next step in the procedure. You may want to have other people help you without them knowing what your project is and without them knowing anything about functions. As you move through the steps, insert appropriate functions into the questions and ask other people to answer those questions for you. Eventually you will reach Step 7, which is the step that tells you how to start the development of your FAST Tree. This is the second time you have the opportunity to have friends help you with your project. Just insert two or three functions into the Why-How Logic Questions and ask them to answer those questions for you. Make sure you answer the questions yourself.

Next, convert your answers and any answers given by other people to functions. Repeat the process each time you have any new functions. A chat room, e-mail, correspondence, a visit, or a telephone call is an excellent method of having other people help you with your Individual Logic Diagram. This is especially helpful whenever you name a function that you know very little about. I frequently have problems with my computer, so I use this approach when I call my son to help solve my computer problems. He has a degree in computer science and always seems to know the correct answer. Once you know how to perform each step, try developing several logic diagrams on different subjects of interest. After you have accomplished this, you should have the skills required to act as a team leader in developing diagrams with the help of other participants.

COMPUTER TEAMS

There are at least two ways of performing team analysis using the computer. The first task required when performing either of these team analysis approaches is to select several participants to assist you in creating a logic diagram for a given project. These people should, of course, have their own computers. They can reside anyplace in the world and should understand FAST and know how to apply the various steps and techniques. The next task is to decide if you are going to develop a Step-by-Step Logic Diagram, which means you are going to do each step and merge your results with other participants before you start the next step in the procedure or you are going to develop a Composite Merged Diagram wherein all participants develop their own diagrams and then merge the results into a single diagram.

The people involved must have the capability to connect to the Internet so all participants can communicate electronically with each other. If you initiate the project, then you should be the team leader. All correspondence must be funneled through the leader and distributed to the entire team. Two or three team members who are skilled in FAST can be very productive. A maximum of five people is recommended for any given team.

These two methods are especially useful for governmental agencies and corporations with personnel in several branches or divisions in different locations who frequently meet to coordinate and solve various problems. A common problem can be solved by a creative solution in a fraction of the time it could be solved by other methods and with minimum expense. The FAST technique not only solves a problem faster, but also eliminates travel, lodging, and meal expenses; filling out expense reports; time lost in making travel, lodging, and conference room arrangements; and layovers, traveling to and from destinations, writing trip reports, and justifying making the trip in the first place, among many other hidden expenses, including tips and other charges.

CREATING STEP-BY-STEP DIAGRAMS

The step-by-step approach starts by conveying a few facts about the selected project. Then each participant names as many functions as he or she can by a certain date or within a given time limit. The functions are then shared and discussed until a consensus is reached. The next step in the FAST procedure is then performed. This activity is then shared, different points of view are considered, and discussions and communication take place until agreement in semantics of function names occurs. A chat room may be the best place to exchange thoughts, ideas, and concepts. At the close of each step, any new functions named or identified are shared with the other participants. Each step of the FAST procedure should be followed and agreement reached by all participants before proceeding to the next step.

This process is continued until each participant has completed Step 6, which identifies the basic function of the project. As the team members start Step 7, they should generate their own lists of available functions and share them until all participants have the same list. The Why-How Logic is then used as each participant develops primary path functions within his or her own FAST Tree. Functions included within the primary path should be identified with a "7" in the list of functions. Participants should not limit their thinking to just the functions listed. The same procedure should be followed in performing Steps 8, 9, and 10. All differences identified in the specific performance of each function should be resolved before proceeding to the next step.

The naming of functions, the verifying of the correctness of the logic, and the opportunity to creatively and collectively explore other methods of performing each function can take place as the logic diagram grows. Steps 11, 12, and 13 may be considered if the FAST Tree has not yet yielded the desired results, but they are not always essential. When all the steps have been completed, the final merged logic diagram then represents the input of all participants. It is usually desirable to convert the FAST Tree into a FAST Diagram once the final tree has been created. The only drawback to this step-by-step approach is the fact that some participants wait for others who are more forward and convincing to control the logic and thinking as they progress. More in-depth thinking is likely to occur when the composite merged approach is pursued. See the paragraph on right of ownership later in this chapter before proceeding with any given approach.

COMPOSITE MERGED DIAGRAMS

The Composite Merged Diagram approach offers the greatest opportunity for independent thinking and creativity. The first three steps for creating this diagram are identical to the Step-by-Step Diagram. The next eight steps are performed by each participant individually. The depth of thinking and the logic that accompany these steps disclose each person's ability to be creative. Participants are not influenced by other participants' thinking, comments, logic, function naming, etc. They are free to role-play and contemplate how the other participants might be approaching the same initial facts. The good thing about using the computer and working alone is that each participant can keep a record of each step as he or she names functions, asks the Why-How Logic Questions, and verifies if the logic holds in both directions and any other information generated or collected while reasoning things out within one's own mind. Participants should always periodically copy the logic diagrams they compile and paste them at the end of their computer files before modifying them. This method of copying, pasting, and modifying what was generated earlier is demonstrated as each of the project steps are analyzed in Chapters 8 though 12.

It soon becomes evident that an individual's education, experience, vocation, surroundings, lifestyle, moral and religious values, peers, relatives, cultural customs, teachers, mentors, as well as successes and failures in life all influence the logic diagram one creates. As participants engage in asking the How Else Question of various functions, their ability to role-play, be creative, and use their imagination is also displayed in their final logic diagrams. It is recommended that the FAST Tree method of documenting the functions be used to create the logic diagram. This permits the logic diagram to grow vertically by adding additional lines and moving functions to the right or left using tab spaces in order to maintain the proper logic. A legend is sometimes placed at the bottom of a FAST Tree so anyone can properly read and understand its content. The legends may vary, especially when more than one method of performing a function is represented by symbols. A sample of a typical legend is shown in Figure 6.1.

Why functions are posted one tab space to the left and above the function being investigated.

How functions are posted one tab space to the right and below the function being investigated.

Supporting functions are posted directly below the function they support preceded by a caret (^) symbol.

The *basic function* to be performed is the top function in this FAST Tree.

Figure 6.1. Typical FAST Tree Legend

This FAST Tree method of developing a Composite Merged Diagram requires that each participant be given only the initial facts about the subject. If the project is an assembly, then names should be given to each part so some continuity will exist among the logic diagrams. The first three steps of the procedure outlined above for creating Step-by-Step Diagrams should be completed. Then each participant is to complete the next eight procedural steps individually without any communication with other team members. A participant may, however, contact the team leader when necessary and submit his or her completed FAST Tree after completing Step 11.

Once the individual logic diagrams have been created, all participants collectively merge their diagrams into a Composite Logic Diagram. The merging of several logic diagrams pertaining to the same subject offers the possibility of opening many doors to understanding, creativity, and opportunity to communicate that would not otherwise be considered. When people work alone and complete logic diagrams by themselves, the diagrams not only reflect individual points of view but also reflect their individual skill in naming functions and answering the Why-How Logic Questions. They are not influenced by how other participants answer these same questions, as they would be if the step-by-step method were to be used. In addition, when people work alone, they have a tendency to think more deeply than they would otherwise. We all have a tendency to follow the person who seems to have the right answers rather than do our own thinking, especially if we are quiet and somewhat reserved. It's just easier that way. I frequently function that way myself because sometimes I am too lazy to think more deeply.

Once all participants have developed their own FAST Tree or FAST Diagram, copies of each participant's work are sent or transmitted to all other participants by the team leader. Then each participant should try to merge all the trees or diagrams together in some fashion. As the participants do this, they will discover many new functions which they did not consider. Also, they may give certain functions an interpretation which is different from what was intended by the originator. This almost always sparks new opportunities for creativity and understanding.

It may take some time to merge the information supplied from other participants. At this stage of the development, intercommunication among all parties is encouraged until some semblance of agreement is achieved. Keep in mind that many times functions are defined differently but often mean the same thing. Don't spend time nit-picking.

Many new avenues will become evident as each participant merges the diagrams together. All merged diagrams are then shared, and the team comes up with the final Composite Logic Diagram. Normally, two or three people still control how the thinking trend goes, even on composite diagrams. The input of every individual at least must be considered in the final diagram. It becomes a unified effort to merge all the diagrams into one which all can agree upon. It also allows all the team members to expand their understanding as they merge the various branches of their logic diagrams together.

Participants should feel free to contact other participants whenever they do not understand someone else's thinking process or the meaning of a particular function. Once again, as others explain how they perform a given function, record what you are thinking. When they get through with their explanations, you may discover that what you visualized in your mind is completely different from what they intended to convey. If your concept is better, then write it down. Also, just because the wording you use is different does not mean one is right and the other is wrong.

There is no reason to continually argue about the wording of a given function after everyone has expressed their belief about it. Just accept the name given to it and mentally tell yourself that it means the same as your point of view. If, perchance, the discussion discloses another point of view that is completely different, use whatever information you glean to further expand your understanding. The final two steps for a merged diagram are performed by all participants working together.

RIGHT OF OWNERSHIP

When you ask other people to become members of your analysis team, you run the risk of someone within the group feeling that they own a particular idea because they recognized the idea first. Team members should realize that the questions, discussions, and communications among them as well as other factors are the ingredients that allow them to conceive a particular concept. The best way to avoid any conflicts is to agree beforehand that the entire group owns any new ideas jointly. This type of agreement encourages full participation. If you are being paid to conduct a study and the other participants are being paid as well, then whoever is paying the bill should own the ideas. When people are being trained to learn FAST and pay a fee, they generally forfeit their right to ownership. Most projects selected for training purposes are usually used with the understanding that the benefits and any new concepts belong to the owners of those projects. Joint ownership of all new concepts ensures that creativity is not inhibited.

NORMAL LOGIC DIAGRAMS

Normal Logic Diagrams usually are created by several participants located in the same room. It is common to have several teams work on the identical project or part of a larger project in the same room. Each team generally consists of five participants: the owner or his or her representative, who is the most knowledgeable about the project; the team leader, who knows and understands FAST and sees to it that the participants stick to the rules or steps of FAST; a specialist who is knowledgeable about the product or service being investigated; a consumer or possible user of the product or service; and it may be beneficial at times to select a participant you respect for having good common sense. Hopefully, one of these five people is you. If not, then include yourself and have a team of six people.

Frequently, other people are consulted at different stages of the analysis, especially when no one on the team has the expertise needed to answer technical or monetary questions. Occasionally, the names of some functions are beyond the comprehension of those assigned to a team. In those cases, specialists may be asked to explain in great detail what a particular function means, why it is required, and how it is presently being accomplished.

Normal Logic Diagrams are generally created as FAST Diagrams rather than as FAST Trees. The steps of the procedure are the same; however, the diagram is constructed on a flat surface instead of on a computer. The room setting should create an environment where the FAST Diagram can be used readily by everyone, and everyone should concentrate on the same function as it is posted or placed on a flat surface such as a wall, a table, or perhaps a flip chart. This allows the participants to reason and determine within their own minds if they agree with the logic other people are proposing. I like each person on the team to answer every logic question if they can. In other words, I like to pick their brains and force them to think more deeply. I also demand answers from myself written as a function for each question I ask.

This approach is similar to the step-by-step approach when using the computer. The major difference is that the Normal Logic Diagram approach allows some participants to just be observers unless the team leader forces them to participate. As the logic diagram grows, all participants should ask and answer the logic questions. If any disagreement exists in the naming of a function or the logic associated with it, it should be resolved before proceeding with the development of the diagram. In a room where all participants are working together, those who are more forward and have greater expertise tend to dominate what happens.

The whole idea of this technique in this setting is, with the aid of other participants, to increase your understanding of a somewhat complicated concept. What actually happens is that something that appeared to be straightforward suddenly becomes more complicated. This occurs because of all the new functions that raise their heads as the process progresses. Be patient — it all turns out better in the long run than it would otherwise. Many times, the team discovers that the real problem was not the one it started with because of the input from all the participants. The participants' education, experience, background, and accomplishments as well as defeats all contribute to the communication that takes place as the logic diagram is formulated.

SELECTING TEAM MEMBERS

A typical team for constructing a logic diagram for a *consumer-type product* consists of five decision makers from within the manufacturing organization. These participants are usually supplied from:

- Engineering
- Manufacturing engineering
- Purchasing
- Estimating
- Marketing

When analyzing *manufacturing procedures,* consider using:

- The person responsible for the manufacturing process
- A worker who is capable of performing almost any assignment

- An industrial engineer
- A foreman
- A product engineer for a high-volume product

Other areas that can be analyzed (such as schools; local, state, and federal governmental agencies and services; religious and community projects) all require judgment in selecting participants. Keep in mind that no one wants to change the way things are presently performed or accomplished unless they are convinced it will be better than what they have already agreed to or what they feel comfortable with. This is especially true when the project involved was designed or created by a specific person or two or three people. The best way to gain their acceptance is to involve these people in the logic thinking exercise.

An excellent treatise on selecting team members and teamwork appears in *Value Engineering: A Plan for Invention*[1] by Richard J. Park. Chapter 10, entitled "Teams and Teamwork: A Synthesized Knowledge Group," covers teams for overall understanding, population and ability, creative-action distribution, team organization, team development, nontask team issues, and how to become a team. Mr. Park also includes an excellent treatise on creativity in Chapter 11 that is interesting and beneficial.

MULTITEAM ANALYSIS

When two or more teams are given the same project or different segments of a major project, interaction among team members takes place as the project progresses and again when all the teams come together to finalize their assigned portions of the total project. Even early in the project, when the initial functions are being formulated, members may disagree as to the wording of a given function. Obviously, when this happens, a discussion takes place, which is usually very beneficial. As participants respond to the Why-How Logic Questions, it is a good idea to have each participant give an answer that is different from any answer given earlier to the same question. The answers are usually general in nature rather than specific. The more general the answer when converted to functions, the greater the opportunity to stimulate creativity. This subject will be treated by showing several examples in Chapter 13.

As the teams merge their FAST Trees or FAST Diagrams together on large projects that have several teams covering different areas of the project, they are able to discuss each of the functions as a group; therefore, many different opinions usually are expressed. Care should be taken to ensure that everyone has an opportunity to express what they think about the naming of the functions and the correctness of the Why-How Logic. This increases the communication among the participants and broadens their understanding. Keep in mind that you have to be practical about what you are doing. Many times, you will want to limit your discussion to a certain branch of the logic diagram. When this happens, try to use some of the earlier steps to creatively expand that portion of the tree you select to pursue.

NOTE

1. Park, Richard J., *Value Engineering: A Plan for Invention,* St. Lucia Press, New York, 1999, pp. 157–189.

7

INTUITIVE LOGIC

Intuitive logic is enjoyed by everyone. You have it whether you realize it or not. I enjoyed intuitive logic for years without knowing it. We all use it every day.

INTUITIVE THINKING

Intuitive logic is the thinking and logic reasoning that take place within a person's mind whenever he or she intuitively plays a role to answer any question. It is also the logic we usually use when we answer the Why-How Logic Questions and likewise as we analyze and evaluate our answers which have been expressed as functions.

At the end of Chapter 3, you were asked to pretend you were teaching someone else as you were learning how to name functions. You were given six different projects to work on. As you did this, you were actually playing the role of the instructor or teacher for those projects. That very act was a form of creativity. If I had not suggested that you play the role of instructor, you probably would have intuitively selected the role of instructor on your own.

We all play several roles each week without realizing it. See if you can list six roles you normally play each week:

_____   _____   _____

_____   _____   _____

As you attempt to do this exercise, think of all the different things you do each weekday. A few that come to mind for me are husband, engineer, father, tutor, teacher, driver, repairman, plumber, mechanic, fire builder, and cook. I know my wife would put a big question mark next to cook. She probably still remembers the first time I cooked the evening meal shortly after we were married. I didn't know you have to boil gravy. I thought all you had to do was mix flour with water and warm it up. Homemade paste really makes potatoes taste awful.

The best thing about role-playing is that it usually occurs intuitively. By that I mean you switch roles automatically without consciously realizing it has happened. For example, suppose you are a passenger riding in an automobile, enjoying the homes and the beautifully landscaped yards as you ride along in a residential area. Or imagine you are riding down a canyon road while someone else is driving as you enjoy the mountain stream and scenery. What happens when you are looking out the window and suddenly you see a child or a deer crossing the road a short distance in front of the vehicle? How do you react? Don't you immediately react as though you were the driver and pretend you are pushing on the brake or turning the steering wheel?

INTUITIVE ROLE-PLAYING

This type of reaction is intuitive, and you instantly become the driver for that short moment. You do the same thing when someone asks you a question about a function. If I ask you how to *catch fish,* what do you think about immediately? What role do you play when you hear that question? Is it the same if I ask how to *wash a car*? No! Your mind suddenly goes somewhere else, doesn't it? The function "catch fish" probably causes you to imagine you are in a boat or beside a cool mountain stream, while "wash car" causes you to imagine you are in your driveway with a hose and a sponge in your hands or at the car wash with a wand in your hand to spray, soap, and wash the grime off your car. You don't have to decide what role you are going to play. You just do it intuitively.

The trick in this technique is to see if you can train yourself to switch roles intuitively every time you ask the same question about the same function. Try doing this whenever you are unable to come up with a good verb-noun function regarding a question about a function. Keep asking the question over and over again. Suddenly, while you are asking "how" of "catch fish," you will be thinking about a tuna schooner or a fishing boat because you are concentrating on the following How Question:

How can **catch fish** be accomplished or performed?

This question will cause you to initially switch to a role you played sometime during your lifetime when you were fishing, but if you concentrate on obtaining a different answer when you ask the same question again, the switching to a new role will begin to happen. At least it did for me without realizing it. Shortly after I learned about functions and started to ask how a particular function is accomplished or proposed to be accomplished, I wrote an article on creativity to use in the seminars I was conducting. I found the following quotation from the great American inventor Charles Kettering, who was head of research at General Motors. It was so inspirational that I immediately adopted his method of looking at any problem or project.

CHARLES KETTERING'S RULE

It's a very simple rule and the only rule we have. It is just this: "The Job Is The Boss." What does the engineer think of this new piston? That doesn't matter. What does the engine think about it? That does matter. The engineer's opinion is worth very little. The engine's opinion is worth a great deal.

If the engine says, "I like this piston," and it happens to be contrary to the engineer's pet idea, that's too bad. It simply proves that the engineer was wrong. After all, as we said, the only reason for all this expensive research is that it corrects our ignorance factor so that we can see the problem in its true light.[1]

Charles Kettering said that sometimes you have to imagine yourself as a molecule of gasoline inside that combustion chamber and visualize yourself going through what that molecule experiences as it is compressed, ignited, and explodes. I suppose I adopted his philosophy as I asked questions about functions. Whenever I asked a question and failed to get an answer that increased my knowledge or stimulated my thinking or creativity, I repeated the question over and over again until I obtained an answer that stimulated me. Invariably, someone would give me an answer or I would think of one that broadened my understanding. I continually repeated the question until I obtained new information or until something sparked my creativity. Every time I attempted to teach seminar participants *why* they wanted to perform a given function or *how* they thought a given function should be performed or proposed to be performed, I used this approach.

Seven years after I presented my first paper on FAST, I found out what I was doing without knowing it. I was teaching a group of participants how to ask the Why-How Logic Questions and had selected the problem of airline highjacking for discussion. That was long before there were any security measures at our airports. This realization came to me in Troy, Michigan at the Society of American Value Engineering Regional Conference, which was organized by Richard J. Park of Birmingham, Michigan. This conference was dedicated entirely to FAST diagramming.

The function I decided we should investigate was "prevent highjacking." As I proceeded to ask why we wanted to prevent highjacking, the participants gave me various answers, which I quickly evaluated and then rejected when they failed to motivate my thinking. When I didn't get any information that increased my understanding of the subject or stimulated my creativity, I repeated the same question again, two or three times. Finally, when someone gave me the same answer that had been given earlier, I said with excitement, "That's it!" A young lady by the name of Donna Rogers said to me, "How come you like that answer now and you didn't like it a few minutes ago?" I responded, "I'm president of the airline now. I was the pilot and a passenger before." It was not until that very instant that I realized I was switching roles each time a particular role failed to yield an answer that increased my level of understanding or to stimulate my creativity.

SWITCHING ROLES INTUITIVELY

If Donna Rogers had not asked that question, I do not believe I would have discovered that I was changing roles. I still remember wondering at the time if I had several personalities. How could I have been switching roles for so long without realizing it? Whenever you get a stroke of inspiration and want to evaluate what inspires you, perhaps you too are evaluating it from the point of view of the intuitive role you are playing, as I was doing.

When a product is involved, the role most often intuitively selected is a consumer. A few people intuitively select a role closely associated with their responsibility, such as manager, designer, engineer, or instructor. When several participants, who have different experiences, education, and skills, are asked a question about a given function, each one of them intuitively will select a role as they attempt to answer the question. Generally, if four or five people are participating, two or three of them will select different roles. Therefore, their answers, expressed as functions, may be somewhat different. If this happens, several new functions and/or ideas may come to light, which will immediately spark a lively discussion. The naming of a new function always opens new doors because the procedure then suggests that we ask the Why-How Logic Questions of those new functions. Any time a difference of opinion exists, it seems to bring to light more information relating to the subject being discussed. At first, everyone thinks they know everything about a well-known subject. As they continue to ask the Why-How Logic Questions, they soon realize that no person knows everything about the project. This is when the group begins to meld into one unified team of cooperative participants.

If you intentionally divorce yourself from the role you have been playing and select a different role to play, you immediately stimulate additional creativity. The best way to do this is to brainstorm a list of functions using the How Else Question. Play as many different roles as you can. You need to remember that whenever you answer the How Else Question, you essentially are describing how that function is performed for the role you are playing. As soon as you change roles, the details as to how the function is to be accomplished could be completely different. *It may be that intuitive logic and the various roles we intuitively select are the reason why FAST is considered to be an excellent communication technique.*

If I ask you why you visit Aunt Hazel, you intuitively start thinking about all the visits you have made to Aunt Hazel's home. Maybe it's because she offers you a piece of pie every time you visit her. Now if I ask you why you visit relatives, you intuitively start thinking about your blood relatives and then about your companion's blood relatives in order to respond to my request for an answer. If you are working with a task force team analyzing a project and one of the functions is "visit relatives," every person on the team intuitively will start thinking the same way you do, but their relatives are entirely different from your relatives and their association with each one of them is different; therefore, who knows what answers they will give to the same question.

IN-DEPTH THINKING

It is hard work to force yourself to think deeply. Most of us just do normal routine thinking out of habit. When you are using this technique, you have to force yourself to perform in-depth thinking, especially when you ask the Why-How Logic Questions. You have to try to transform yourself into that "molecule of gasoline," like Charles Kettering used to tell his research engineers at General Motors. By the same token, there is no mental motivation to intuitively switch into a different role unless you are intently thinking and concentrating deeply about obtaining a function you have not thought of before. At least that's the way it works for me.

I will play several different roles as I demonstrate the various steps of this technique in the chapters that follow. I will intentionally identify these roles after I obtain an answer so you will understand how I am able to come up with the answers I record. If I did not disclose the role I am playing, you would have no clue how I came up with the answer I documented in some cases. But as soon as I tell you what role I play, the answer becomes obvious. As I stated earlier, I just keep asking questions about the same functions over and over again and intuitively switch roles until I get results. When they cannot think of a new function, most people conclude that it is best to just move on. They miss many opportunities for creative development when they do this. My students used to ask me, "What answer do you want?" I would say to them, "I'm trying to get you to do your own thinking." If I continued without their participation, I could usually come up with another new function. When this happened, I discovered that I was mentally exhausted after a day's work because deep thinking takes effort and persistence.

You can always imagine yourself being someone else or an item or a thing, as Charles Kettering suggested, before you try to answer a given question. Sometimes I consciously use his approach myself, but generally I just try answering the question and switch roles intuitively. When you are working on a project with several different people, every function causes each person to intuitively switch to the role that matches his or her lifestyle. If the people participating are not responding, you might suggest different roles they might consider.

If you ask a group of four or five people how someone learns honesty, each person may describe the process differently. But if you also ask them to agree on the best way to learn honesty and to express it using just a verb and a noun, what would happen? First of all, each person would tell how he or she learned to be honest. Then a discussion would continue until agreement and meaning of the verb-noun terminology for the combined answer was reached. The same thing should happen every time you ask a question about any function. Intuitive role-playing combined with a discussion among several people always increases the information available for a particular function. It causes various points of view to be considered and enhances communication among all members of the task force team.

As a person approaches any project, he or she intuitively will select a role to play. The role selected may be himself or herself or a companion, boss, neighbor, or best friend. It might be a doctor, a lawyer, or an engineer. A person may even switch roles in trying to arrive at a worthwhile function

We all perform some degree of intuitive logic and role-playing in our daily lives. Since these two elements play such an important part in expanding a person's understanding whenever a project is analyzed, this entire chapter is devoted to this subject. Then the main steps in the construction of any logic diagram will be outlined. After they are covered, you will be taught some other creative innovations.

THE CHIMNEY STORY

Several years ago, I told the following story. I think it is worth repeating here. When my wife gave birth to our fourth child, we felt that we should have a larger home. My father-in-law had agreed to give us a lot just north of his milking shed, which had a large barn next to it. When the mason was laying the bricks, we had him build the chimney just high enough so we could put the roof on before it snowed. The next spring, I decided to finish the chimney myself, so, being the good engineer I thought I was, I got my slide rule out and did some calculating. This was before we had calculators and computers.

After the bricks arrived, I fetched from my father-in-law's barn a ladder that was long enough to reach the roof. Then I proceeded to carry all of the bricks up onto the roof in a bucket, a few at a time. Finally, they were neatly stacked on the roof. Every weekend I would mix a batch of mortar and work on the chimney. Finally I finished the task. However, I discovered that I had miscalculated the number of bricks required. I had ordered twice as many as I needed to complete the job.

Now the problem I was faced with was how to get the bricks off the roof without breaking any of them, because I wanted to take them back to the brickyard and get a refund. I decided the best way to accomplish this was to set up an A-frame structure up on the roof, attach a pulley to it, and then thread a rope through the pulley. I figured that it would take me forever to get all the bricks down off the roof with the small bucket I had used to carry them up, so I went over to my father-in-law's barn and found an old wooden barrel. I tied a three-legged yoke to the top of the barrel and attached it to one end of the rope. Then I raised the barrel just above the eaves of the roof of the house and tied the other end to a railing near the ground. Next I climbed the ladder and proceeded to fill the barrel with bricks.

When the barrel was almost full, I climbed down the long ladder and untied the rope. That is when I made an important discovery. I discovered that the barrel of bricks was slightly heavier than I was. I hung on to the rope, and as I was going up I met the barrel of bricks coming down, which hit me on the shoulder as we passed each other, but I hung on. When I arrived above the eaves of the house, my fingers got caught in the pulley, and boy did that hurt. Just then, the barrel hit the ground, and its wooden bottom broke out, leaving the bricks in a pile on the ground. Now I was heavier than the empty barrel, and as I went back down, I ran into the empty barrel as it passed me going back up. When I hit the ground, I landed on top of the pile of bricks. That is when I realized what had happened and decided I should let go of the rope. A moment or two later, the empty barrel came down and hit me on top of my head.

I just have one question for you. Did you intuitively imagine yourself being raised off the ground as I described the events that took place as soon as I untied the rope? Always remember what Albert Einstein said: "Imagination is more important than knowledge!" Use your imagination every chance you get.

ROLE-PLAYING: A LIFETIME OF EXPERIENCES

The role-playing technique works equally well for each member of the task force team when asked the Why-How Logic Questions concerning a given function. Team members have different skills, education, talents, experiences, associates, companions, cultural customs and traditions, family, schoolmates, friends, neighbors, political and religious leaning, etc., and their individual intuitiveness will take them in many different directions. What has happened to a person throughout his or her whole lifetime will influence the response to these two logic questions. Each team member should be asked these questions. Then, "the team members must dialogue and reconfigure the FAST model (*diagram*) until a consensus is reached and all participating team members are satisfied that their concerns are expressed in the model" (italics added).[2] After I presented a paper in 1992 entitled "An Intuitive Thinking Technique" at the SAVE International Conference, Jerry Kaufman of Houston, Texas named this thinking exercise *intuitive logic.* I liked the term intuitive logic because it combined my Why-How Logic and my intuitive thinking and role-playing into a single term; therefore, I named this chapter "Intuitive Logic."

When several people are asking questions about functions, rarely do they know what roles are being played by the other people participating. This causes disagreements as these people answer various questions. This type of disagreement forces the participants to discuss their answers in greater detail. As they do this, each person has a tendency to adapt what is said to the role he or she is playing. This is the ideal situation for expanding understanding and stimulating creativity. It allows the team to explore all facets of a problem or subject in just a few minutes, which otherwise might take hours to accomplish. The same type of imagination you experienced as you read my chimney story takes place as you explain how or why you perform a given function. Your imagination fills in the blanks that are missing and frequently allows your thinking to visualize a concept that is better than what was being explained. This also is a method of stimulating creativity.

GETTING OTHERS TO PARTICIPATE

An excellent way to get other people to aid you in your creative endeavor is to supply them with a list of functions and have them sit at their own computers and answer the Why-How Logic Questions for each function using a word-processing program. This is discussed briefly in other chapters within this book. Just send them your list of functions over the Internet or by regular mail, along with the following two questions:

How is *this function* actually accomplished?

Why must *this function* be performed?

Tell them to insert each of your functions into these two questions where *this function* appears and give you the best answers they can think of. Tell them you are trying to teach them how to think FAST or do FAST diagramming. Their curiosity will probably motivate them to comply with your request. You never know what points of view people will take when they study your questions, especially when you do not disclose what project you are working on. They invariably will intuitively pick a role that will disclose avenues you would not otherwise consider.

PARTICIPANTS SHOULD BE TRAINED

One of the problems new participants have when asked to work on a given project is that they often become negative before they learn the basic concepts of FAST. Therefore, they should be trained in the basic concepts before they work on a project. Years ago, while I was conducting a seminar at Sperry Phoenix, a division of the then Sperry Rand Corporation, a gyroscope engineer who was assigned to work with our team got frustrated after one day of training and asked to be taken off the project. I think his frustration occurred because I told him we were going to develop some new concepts in gyroscope design, to which he responded: "Do you know that there are only about a hundred people in the entire world who really understand gyroscopes? And you think we're going to develop something new?" I responded by saying, "That's right!" I had to admit to him at the time that I really didn't know how gyroscopes worked.

The next day, I was asked by his boss if I would accept a replacement on the team. I responded by saying, "We have this fellow half oriented already, and he doesn't even know it." While our team was brainstorming and asking the Why-How Logic Questions of functions we had identified, this engineer walked into our meeting and started complaining because his boss had told him he had to remain a member of our task force. He said, "I guess I'm forced to go along with your silly game." Within a half hour, this engineer was up at the board modifying someone else's concept. He became so enthusiastic about this new concept that he devoted his own time after working hours and made drawings of what we had discovered. This new concept, along with five other concepts which had never before been considered for improving gyroscopes, was presented several weeks later to management. Sperry Phoenix's management would not allow me to publish an article on our results because the concepts were considered to be proprietary information.

If you have several people assist you on a given project and are able to get them to respond to your Why-How Logic Questions, they intuitively will pick a role and give you answers to your questions that are in agreement with that role. Once these roles have been activated, the opportunity for creativity is at its highest level. *It is a known fact that rarely does the listener receive the exact same concept the speaker intends to send.* You see, if I become creatively stimulated by a particular function, then I should be the one who tells how to accomplish that particular function. If someone else is creatively motivated, then they should tell how to perform that function.

When I describe how to perform a function in the role I am playing, other members of the team adapt what I say to the role they are playing, just like you did as you read my chimney story. You see, whenever we tell someone else what we are thinking, we always leave out a few important facts. I failed to mention in my chimney story that the new house I was building was only one-story high. If I had told you that fact, your imagination would not have been very active. People think they know what you are saying, but if they are playing a different role than the one you are playing, what you say doesn't quite fit. Therefore, they modify your concept so it does fit. Sometimes their concept is entirely different from the one being described and often is much better.

EVERY PERSON SHOULD PARTICIPATE

Each team member should, in turn, be asked to explain how he or she thinks you answered a particular question or explain a concept you presented. When this occurs, you and the other team members now adapt what this person says to the various roles the rest of the team members are still playing. The end result may be the development of several new ideas or concepts. At the very least, a broader understanding of the subject will be acquired.

If a member of your team fails to respond with any suggestions or with his or her own description of how he or she would perform a given function, ask that individual to explain one of the concepts presented by someone else. This request enables you to capitalize on every member's intuitive role and creative abilities. Invariably, people will always add to or modify what was presented initially. Hitchhiking and modifying concepts can, within minutes, yield several new concepts and ideas.

This reminds me of a story I heard at a friend's funeral. The fellow who died liked to tell this story. A lady in the community was known to gossip a lot. One day, several neighbors were talking about various things and people in general who lived in the neighborhood. One of the ladies proceeded to tell the group what she had heard about a young lady who had recently moved into the neighborhood. The more she talked, the more the people wanted to know. They kept asking her to tell them more, time after time. Finally, she responded by saying, "I've already told you more than I heard!" We want participants to tell everything they mentally imagined after listening to someone else's concept.

OPENING THE LINES OF COMMUNICATION

By computerizing the activities performed while applying FAST Creativity, several people can each take the same problem and within a very short period of time create their own lists of initial functions along with basic functions, why functions, and how functions on their own computers. These functions can then be transmitted to other participants by e-mail. Then each participant can compare his or her functions with those generated by other participants, thereby opening the lines of communication even more than when all team members are in the same room. In a room where all participants are working together, those who are more forward and have greater expertise tend to dominate every discussion. Generally, two or three people control how the thinking trend goes, while the others just listen or observe. If, on the other hand, these same people initially work alone on their own computers, they are forced to do their own thinking and they think much more deeply because they know what they are recording is going to be transmitted to other participants.

Also, by using the computer, each person is able to intuitively select a role and thereby contribute perhaps a different point of view, thus opening avenues that would not otherwise be considered. Generally, the quieter a person is, the more thought provoking his or her thought process is. A chat room on the Internet also can be used by several individuals who are miles apart or perhaps on different continents, as discussed in Chapter 6. The computerized approach uses the same basic concepts I presented in my initial paper, but it is now reinforced with a couple of new questions, recommended procedural steps, examples of how one should think and analyze, how to role-play, how to write how functions without disclosing the method of accomplishment, how to use a thesaurus, and several other demonstrated approaches. These aids should help each participant as his or her thought process expands from the initial function level to the higher and lower function levels.

NOTES

1. Bytheway, Charles W., "FAST Diagrams for Creative Function Analysis," SAVE Communications and *Journal of Value Engineering,* Vol. 71-3, March 1971, p. 9.
2. Kaufman, J., *The Principles and Applications of Function Analysis System Technique FAST,* J. Kaufman Associates, Inc., Houston, Texas, 1999, p. 39.

8

PROJECT 1:
LIGHTBULB

Up to this point, we have talked about a number of things that pertain to FAST, but we have not established a procedure to follow. We will now talk about the first 11 steps of the FAST procedure. As we do this, we will first apply the first 10 steps to Project 1, which is an incandescent lightbulb. Then I will show you how to apply Step 11. The main reason for selecting this project is to demonstrate how to construct a logic diagram in the form of a FAST Tree, which was introduced in Chapter 4. Then we will convert the FAST Tree into a FAST Diagram. No attempt is made to be creative as the diagram is being constructed other than naming the initial functions and answering the Why-How Logic Questions.

PROJECT SELECTION

Step 1. Selecting a Project

Step 1 of the procedure is to select a *subject or project* to analyze. Chapter 5 presented five different methods of accomplishing this step and five projects were identified. In this and the next four chapters, the various steps of the FAST procedure are demonstrated for each of these projects. The five methods of selecting a project are listed in Figure 8.1.

The questions to ask when using Methods 3, 4, and 5 are listed in Figures 8.2, 8.3, and 8.4, respectively.

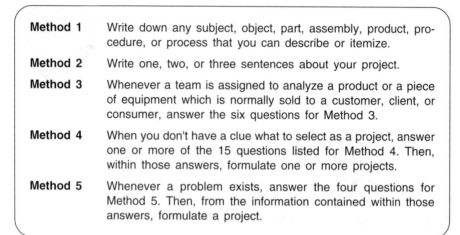

Method 1	Write down any subject, object, part, assembly, product, procedure, or process that you can describe or itemize.
Method 2	Write one, two, or three sentences about your project.
Method 3	Whenever a team is assigned to analyze a product or a piece of equipment which is normally sold to a customer, client, or consumer, answer the six questions for Method 3.
Method 4	When you don't have a clue what to select as a project, answer one or more of the 15 questions listed for Method 4. Then, within those answers, formulate one or more projects.
Method 5	Whenever a problem exists, answer the four questions for Method 5. Then, from the information contained within those answers, formulate a project.

Figure 8.1. Methods for Selecting Projects

Q1	What product or piece of equipment have you been assigned to analyze?
Q2	What is the main purpose for which this product has been built or assembled?
Q3	How can this product be made so it is more dependable?
Q4	How can this product be made so it is more convenient to use?
Q5	How can this product be improved?
Q6	How can this product be made so it is more pleasing to the five senses?

Figure 8.2. Questions for Method 3

Step 1 for Project 1. Selecting a Project

Method 1 suggests that an assembly of some kind be selected. The incandescent lightbulb assembly shown in Figure 8.5 will be used for this project. It has been selected for demonstration purposes to show how the logic helps increase one's understanding and how to add supporting functions to a logic diagram. This project was first presented in 1965.[1]

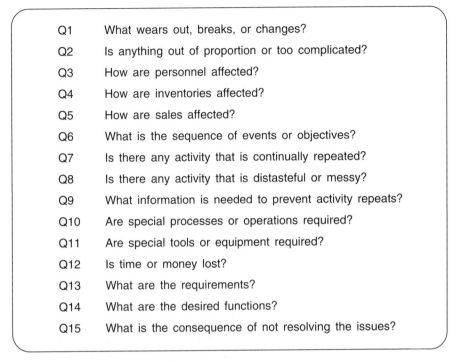

Q1	What wears out, breaks, or changes?
Q2	Is anything out of proportion or too complicated?
Q3	How are personnel affected?
Q4	How are inventories affected?
Q5	How are sales affected?
Q6	What is the sequence of events or objectives?
Q7	Is there any activity that is continually repeated?
Q8	Is there any activity that is distasteful or messy?
Q9	What information is needed to prevent activity repeats?
Q10	Are special processes or operations required?
Q11	Are special tools or equipment required?
Q12	Is time or money lost?
Q13	What are the requirements?
Q14	What are the desired functions?
Q15	What is the consequence of not resolving the issues?

Figure 8.3. Questions for Method 4

Q1	What problem shall we discuss?
Q2	Why do you think this is a problem?
Q3	Why do you think a solution is needed?
Q4	What is there about this problem area that disturbs you?

Figure 8.4. Questions for Method 5

ONLY ONE PARTICIPANT

Step 2. Selecting Participants

Step 2 is the selection of one or more participants to assist in the analysis process. You may elect to skip this step if you choose to do the analysis by yourself when you create an Individual Logic Diagram. A maximum of five

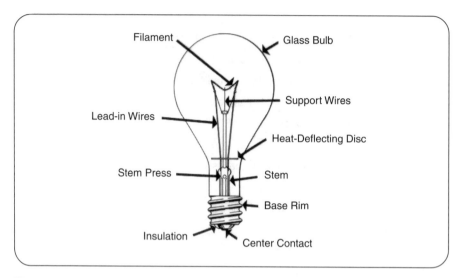

Figure 8.5. Lightbulb Assembly

participants is recommended, if you select any of the other three logic diagram approaches. The skills and talents of those you should select as participants are covered in Chapter 6, along with information concerning creating Step-by-Step Logic Diagrams, Composite Merged Diagrams, and Normal Logic Diagrams. The advantages of having participants use the computer are also covered in that chapter.

Step 2 for Project 1. Selecting Participants to Create a Logic Diagram

This particular project was selected to demonstrate how to create a FAST Diagram when I wrote my first article about this subject. There were no participants other than myself; therefore, an Individual Logic Diagram will be used to demonstrate the next eight steps of this procedure.

CONDUCT CURRENT AND MOUNT LAMP

Step 3. Initial Functions

Step 3 requires you to transpose into functions the statements and facts you obtained when you selected and defined your project. These functions are called *initial functions.* You may end up with only one initial function or perhaps 10 or 20 initial functions, depending upon the project you select. How initial functions are named will be demonstrated as the five sample projects are developed in Chapters 8 through 12. If you invite participants to work with you, have each one develop his or her own list of initial functions. This will ensure that the team gets off to a good start. If either a Step-by-Step Diagram or a Composite Merged Diagram is to be created, each participant should share his or her list of functions with the team leader and receive other participants' lists before performing Step 4. All participants, of course, should know how to name functions. The type of logic diagram to be created should be determined before proceeding to Step 4. The types are individual, step-by-step, composite merged, and normal.

Step 3 for Project 1. Initial Functions

The lightbulb shown in Figure 8.5 consists of a base rim, a glass bulb, some insulation, a stem, a stem press, lead-in wires, support wires, a filament, a heat-deflecting disc, and a center contact. The task is to determine what functions are performed by the major components of this assembly. Figure 8.6 lists the components of an incandescent lightbulb and the functions they perform. As stated earlier, we are creating an Individual Logic Diagram.

PRODUCE LIGHT

Step 4. Initial Basic Function

Step 4 requires you to determine the *initial basic function* from your list of initial functions. The basic function obtained initially is not always the

Parts	Initial Functions
Lightbulb	*Produce Light
Base	Conduct Current
Base	Mount Lamp
Glass Bulb	Exclude Oxygen
Insulation	Insulate Conductors
Filament	*Produce Light
Stem Press	Provide Airtight Seal
Disc	Deflect Heat
Misc. Parts	Position Filament

Figure 8.6. Initial Functions for Lightbulb

function that should be selected to begin the logic diagram because frequently the real problem or project that should be investigated is at a higher level in the logic diagram. Therefore, subsequent steps are used to make sure that the real problem or project is identified before effort is devoted to the logic diagram. If the Basic Function Determination Question yields all "no" answers, then the initial basic function has been found. After you have identified the initial basic function, place an asterisk (*) in front of it in your list of functions. The following Basic Function Determination Question is used to obtain the initial basic function:

> If *this function* didn't have to be performed, would any of the other functions still have to be accomplished or performed?

That function in turn is used to search for higher level functions.

Step 4 for Project 1. Initial Basic Function

Obviously, the purpose of a lightbulb is to "produce light." Therefore, if we insert this function into the Basic Function Determination Question, we obtain:

Q   If we didn't have to *produce light*, would we still have to perform the other functions listed?

A   No, if you look down the list of other functions listed in Figure 8.5. This is the answer you obtain for each function.

The *initial basic function* for Project 1 becomes *"produce light."*

PROVIDE LUMINOUS ENERGY

Step 5. Develop Higher Level Functions

This step requires a diligent search for higher level functions. It is accomplished by asking questions of the initial basic function recorded during Step 4. The initial basic function is inserted in the blanks where the asterisks appear in the following three questions:

1. Why is it necessary to _____*_____?

2. What higher level function caused _____*_____ to come into being?

3. What is really trying to be accomplished when _____*_____
   is performed?

Keep in mind that when these three questions are asked, we are trying to come up with three new functions; therefore, we have to force ourselves to think more deeply about our subject or project. When I do this with a team, I always evaluate the answers the participants give me. If they fail to give me a new function, then I repeat the question again and again and again until each participant gives me at least one new function if at all possible. Each time I do this, I try to think of a new function myself. As I do this step, I switch roles and I ask the participants to also play different roles until we are successful in naming some new functions. In other words, pretend you are someone else or something else, such as a piston or a molecule of gasoline.

As soon as I get at least one new function by asking Question 1, either from myself or from the team, I move on to the second question and repeat the procedure all over again. Then I move on to the third question. Many times, the participants say to me, "Why don't you tell us what function you are looking for so we can move on to the next question?" I try to tell them that it isn't what I want; it is what they should want. I am trying to get them to come up with their own new function by switching roles and thinking more deeply.

The only way that happens is if they force themselves mentally to think of a function that has not been mentioned before. If perchance I don't like someone's answer even though it may be an entirely new function, I ask them to explain the function to me so I can understand it. If I still do not understand it, then I need that person to tell me what role they are playing and why they think it is a good function, so I can evaluate the function by playing the same role they are playing. I take it upon myself, as the facilitator, to evaluate the answers because I am looking for new information that will help me as well as the participants to understand the project better. You should try to do the same thing. If an answer does not inspire you, perhaps it will inspire someone else. If it does, ask them to explain how it inspired them and then move on to the next question. From these three questions, at least one new function should be obtained.

Step 5 for Project 1. Develop Higher Level Functions

Since "produce light" is the initial basic function, we insert it into the three questions. Note that I play three different roles as I attempt to answer these questions and record the answers as sentences.

Q1   Why is it necessary to **produce light**?

Role   Waiter at a restaurant
A1   Very little light is available where I work.

Q2   What higher level function caused ***produce light*** to come into
        being?

Role   Worker at a power plant
A2   We produce electricity so residences have power and lights.

Q3   What is really trying to be accomplished when ***produce light***
        is performed?

Role   Foreman who works a night shift
A3   We are trying to make life more productive and enjoyable
        when the sun's light is not available.

You might be thinking, "Those are stupid answers." They might be, but they have a tendency to expand your thinking into other areas you otherwise would not venture. Now that we have these answers, our next task is to glean from them additional functions. A list appears in Figure 8.7, along with our initial basic function.

MAKE LIFE MORE ENJOYABLE

Step 6. Identifying the Basic Function

At this point in the analysis, you should have at least two new higher level functions. These new functions formulated during Step 5, along with your initial basic function, should be used to create your higher level functions list. From this list, you once again select the function you think is the basic function and

Produce Electricity
Provide Luminous Energy
Increase Productivity
*Make Life More Enjoyable
Produce Light

Figure 8.7. Higher Level Functions

insert that function into the following Basic Function Determination Question to determine if it is indeed the new basic function:

Q   If *this function* didn't have to be performed, would any of the other (higher level) functions still have to be performed?

As soon as the basic function has been determined, move on to Step 7.

Step 6 for Project 1. Identifying the Basic Function

From the list of five functions in Figure 8.7, we select the function that appears to be the highest level. "Make life more enjoyable" appears to be that function. Therefore, we insert it in the Basic Function Determination Question, which reads as follows:

Q   If *make life more enjoyable* didn't have to be performed, would any of the other functions still have to be performed?

A   No

Therefore, *"make life more enjoyable"* is the new *basic function.*

DEVELOP HOW TO MAKE LIFE MORE ENJOYABLE

Step 7. Develop Primary Path Functions

Start this step by listing all the functions you have formulated for your project up to this point. Next start the logic diagram, if you are building a FAST Tree, by setting your tab spacing to about 0.3 inches and by posting the basic function one tab space from the left margin in a word-processing file. Then create the next level of functions by indenting another tab space to the right and below this function and post one or more functions which describe how the basic function is actually accomplished in a present design or how you propose to accomplish it if a design or concept does not already exist. If "prevent cavities" is the basic function, the first two levels of the primary path might look like the functions listed in Figure 8.8.

The same procedure is used when creating a FAST Diagram except functions are written on small rectangular cards and placed on a flat surface. Each lower level function is placed at the right of the function being investigated. To avoid confusion, from this point on I will only outline the procedure for creating

Prevent Cavities
Clean Teeth
Eat Wholesome Foods
Avoid Eating Sweets
Use Fluoride Toothpaste
Rinse Food Particles
Floss Teeth

Figure 8.8. Two-Level Primary Path

a logic diagram for a FAST Tree. Refer to the examples in the various chapters to see how a FAST Tree is converted into a FAST Diagram. Also read Appendix A on constructing FAST Diagrams.

When you include a function in your FAST Tree from your function list created at the beginning of this step, post a "7" near the left margin of that function. The "7" indicates that a particular function has been included during Step 7. Always check your logic in the "why" direction to make sure it is correct. Use the Verification Question if you are not sure the logic is correct. The basic function in this case is considered to be the why function, and any new functions which describe how the why function is accomplished are considered to be how functions. Any function you ask "how" of at other function levels is also considered to be a why function. The answers to that question are all how functions.

This process of asking how and why is repeated for all new functions added to the diagram. Gradually build the next lower level of your FAST Tree by moving down one line and indent one tab space as you add additional lower level functions. If two or more functions are posted at the second level or any other level, you will have to insert a new line between the functions and then indent before typing its lower level functions. You may recall that in Chapter 4 we built the logic tree from the bottom in order to learn the basic elements of the Why-How Logic. By now you should have a good understanding of that logic and how to check it, as well as how to type or paste the functions so the logic agrees in the why-how directions.

The higher you start in a logic diagram, the more productive you will be; therefore, always start with the basic function you identify in Step 6. If you have a lot of functions listed, you should change the tab spacing so each function can be typed on just one line. Continue to build the diagram, but always check to see if any of the other formulated functions you listed at the beginning of this step can be inserted. If in doubt about how to accomplish this, refer to Step 7 of the other projects in Chapters 9 through 12.

Step 7 for Project 1. Develop Primary Path Functions

This step requires all the functions identified thus far to be listed in a word-processing program or written on small cards placed on a flat surface. The list of functions for the lightbulb project is shown in Figure 8.9.

Next, post the basic function at the bottom of your file to start building your FAST Tree, and then ask the following question:

Q   How is *make life more enjoyable* actually accomplished?

Obviously, there are numerous ways to "make life more enjoyable." "Provide luminous energy" is only one way. By brainstorming the basic function, all kinds of ways to make life enjoyable can be thought of, which can be expressed as functions. Any new function we conceive might stimulate our creativity. However, after we have explored our list of new functions, we then have to be practical and select the function for which we want to build a primary path. In making this selection, we need to consider which function will include the project we are analyzing.

Exploring these higher level functions often identifies the real problem. The demonstration projects in the succeeding chapters use a variety of ways to decide which path should be pursued. Many times, the basic function deter-

	Produce Electricity
7	*Provide Luminous Energy
	Increase Productivity
7	~~Make Life More Enjoyable~~
7	Produce Light
7	Conduct Current
	Mount Lamp
	Exclude Oxygen
	Insulate Conductors
	Provide Airtight Seal
	Deflect Heat
	Position Filament

Figure 8.9. List of Functions

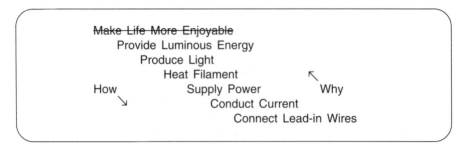

Figure 8.10.   Lightbulb FAST Tree 1

mined during Step 6 is indeed the place where creativity should begin. Other times, it is more practical to select a lower level function as the basic function. Since the manufacturing of a lightbulb was our initial project and was selected for the purpose of showing how to create a logic diagram, we will now select from our list of functions in Figure 8.9 a basic function that will allow us to carry out our initial plan. Therefore, we select "provide luminous energy" as our new basic function. Then we develop the primary path functions shown in Figure 8.10. This is accomplished by asking "why" and "how" of all new functions added. Since "make life more enjoyable" is higher than we elect to analyze, it is crossed out, as shown in Figure 8.10.

Four functions listed in Figure 8.10 appear in Figure 8.9. Therefore, a "7" is added at the left margin to indicate that these functions have been included during Step 7. Also, "make life more enjoyable" is crossed out in Figure 8.9 because we elected to not start at that high of a logic level.

PRODUCE ELECTRICITY AND INCREASE PRODUCTIVITY

Step 8. Evaluate Remaining Formulated Functions

When you are unable to develop your logic diagram further, copy and paste all the functions listed at the beginning of Step 7 that *have not* been included in the logic diagram. Post each function about three tab spaces from the left margin, and separate each entry by three or four lines; then highlight each function by making it bold. Next apply the Why-How Logic to each function posted. As you do this, post your why answers above and one tab space to the left, and post your how answers below and one tab space to the right. The functions thus grouped form a function cluster of a minimum of three functions. This will be demonstrated and made clearer when we perform this step on other sample projects. Each cluster is checked to make sure the logic holds by asking the following question of all how functions:

Does *this **how** function* help *its **why** function*?

If the answer is yes, it indicates that the logic is correct in the how direction. If the answer is no, it indicates that the logic is not correct in the how direction. However, a "no" answer indicates that the how function is a function which supports the why function. When this occurs, a caret (^) symbol before the how function in the cluster is used to indicate that it supports the why function above it. Every supporting function has the potential to be developed into a secondary path of functions. Secondary path functions are also developed by asking the Why-How Logic Questions, just like primary path functions, and have a primary path of their own.

Next, copy the FAST Tree developed during Step 7 and paste it at the end of the working file. Search your function clusters to see if any of the functions can be merged into the primary path developed during Step 7. If any can be merged, insert them in the FAST Tree and place an "8" at the left margin to indicate which functions were merged during Step 8. If a function cluster is at a higher level than you elected to pursue, cross those functions out so you know you have considered them. When this effort yields no more fruitful results, move on to Step 9.

Always remember that every function fits somewhere in a logic diagram. Your task is to find out where. As you pursue this search, your understanding of your project will increase and numerous opportunities to be creative will become available. It is like solving a puzzle; it takes time, thought, and patience to put the diagram together.

Step 8 for Project 1. Evaluate Remaining Formulated Functions

The first thing to do in Step 8 for Project 1 is list all of the functions in Figure 8.9 that have not yet been considered during Step 7. This is accomplished by copying and pasting them at the end of your project file after tabbing in three times for each entry. Separate each entry by three or more lines and then highlight each entry only, and not the entire listing, by making it bold. This allows only the entries to be highlighted when the Why-How Logic Questions are asked of each function. Next develop a function cluster for each function by asking the Why-How Logic Questions and placing your answers where they logically belong using the appropriate tab spaces, as follows:

7	Supply Power
9	**^Produce Electricity**
10	Rotate Armatures

	Make More Products
	Increase Productivity
	Remove Obstacles

7	Connect Lead-in Wires
8	**Mount Lamp** Electrically
8	Twist Lightbulb

9	Position Filament
10	**^Mount Lamp** Mechanically
10	Twist Lightbulb

9	Prevent Filament Oxidation
10	**Exclude Oxygen**
10	Constrain Gases

7	Conduct Current
9	**^Insulate Conductors**
10	Install Insulated Wires

10	Constrain Gases
10	**Provide Airtight Seal**
10	Mold Base to Bulb

9	Reduce Base Temperature
10	**Deflect Heat**
10	Install Reflector

7	Heat Filament
9	**^Position Filament**
10	Position Support Wires

Now check the logic of each cluster. If the Verification Question indicates that the logic does not hold, precede the how function with a caret (^) symbol. This indicates that this function is a supporting function of the function above the caret symbol. These supporting functions are added during Step 9.

Examples of Testing a Few of the Above Clusters

> Q   Does *produce electricity* help *supply power*?
> A   No

This answer is no because the question has to do with supplying power to the lightbulb. You can produce lots of electricity and still not connect it to the lightbulb.

> Q   Does *rotate armature* help *produce electricity*?
> A   Yes

> Q   Does *mount lamp* help *position filament*?
> A   No

This answer is no because the question has to do with the construction of the lightbulb and not positioning the filament within the vacuumed space within the glass bulb. Also note that when the lamp is mounted, it is mounted mechanically as well as electrically.

> Q   Does *insulate conductors* help *conduct current*?
> A   No

> Q   Does *position filament* help *heat filament*?
> A   No

The next task is to copy the FAST Tree in Figure 8.10, paste that copy at the bottom of the working file, and label it Figure 8.11. Then merge as many of these function clusters into the primary path as you can by looking at the

```
        Provide Luminous Energy
           Produce Light
              Convert Energy
                 Heat Filament              ↖
        How         Supply Power            Why
              ↘           Conduct Current
                           Connect Lead-in Wires
                              Mount Lamp Electrically
                                 Twist Lightbulb
```

Figure 8.11. Lightbulb FAST Tree 2

higher level functions of each cluster to see where they might fit. Add an "8" near the left margin of each function within its cluster which you are able to merge during Step 8. Do not add any supporting functions at this time until you are skilled in performing Step 9.

As you build the FAST Tree, be sure to check the Why-How Logic. Note that a new function, "convert energy," has been added. This is an excellent function to brainstorm if we were searching for other methods of converting energy.

INSULATE CONDUCTORS AND POSITION FILAMENT

Step 9. Using When/If Logic to Add Supporting Functions

A supporting function is a function that must be performed when and if a primary path function cannot be performed without this function being performed at the same time or some time prior to the primary path function. For example, a spring that has stored energy within it cannot release that energy if the energy is not stored in the spring to start with. These functions are sometimes identified during Step 8 by placing a caret (^) symbol in front of a function within a function cluster. However, supporting functions usually are identified by inserting each primary path function into the following question:

> When/if *this function* is performed, what other functions must be performed?

If you are unable to think of an answer when you insert each primary path function where "this function" appears in the question, see if any of your developed clusters include a primary path function. If they do, insert just the functions down to and including the supporting functions until you learn how to perform Step 9. Always copy and paste your developed diagram at the end of your file before you begin to add supporting functions. When supporting functions are added during this step, make sure you place a "9" near the left margin of the appropriate function to indicate its inclusion during Step 9. Make sure you consider other functions which have not been named or identified yet. In some cases, a supporting function supports a function in a branch of the diagram other than the primary path. This situation is covered in Step 10.

Step 9 for Project 1. Using When/If Logic to Add Supporting Functions

During Step 8, we identified the following supporting functions within clusters:

Produce electricity
Mount lamp mechanically
Insulate conductors
Position filament

The first task under this step is to see if any of these four functions are supporting functions of any primary path functions by scanning down our list of primary path functions. Since "position filament" is a supporting function of "heat filament," it can be added to the logic diagram during this step by using the caret (^) symbol to indicate that it supports the function immediately above it, as shown in Figure 8.12. Also "produce electricity" can be added below "supply power," and "insulate conductors" can be added below "conduct current." In as much as we are limiting our project to the incandescent lightbulb, we will limit our supporting function investigation to only the functions involved in the manufacture of the lightbulb. As we move to lower level functions in the primary path, we then start again at the top of our primary path list by asking the When/If Logic Question of each primary path function. When we come to "heat filament," we ask:

Q   When/if **heat filament** is performed, what other functions must be performed?

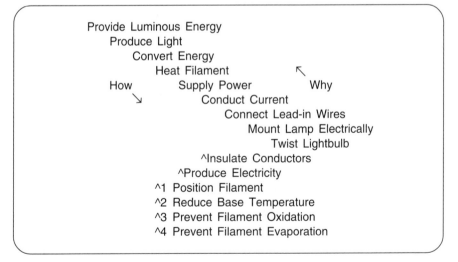

Figure 8.12.   Lightbulb FAST Tree 3

A   When a filament is heated, it can become oxidized if it is exposed to oxygen and can burn up. In some environments, it can vaporize into a gas if the surrounding space is not already occupied by another gas. One way to minimize or prevent these changes from taking place is to control the filament's environment. The answer to this question yields the following three additional supporting functions:

    Reduce base temperature
    Prevent filament oxidation
    Prevent filament evaporation

These three functions are also added to the logic diagram by using the caret symbol. However, since four functions support "heat filament," all four are posted on separate lines and numbered so as to indicate that they all support the same function, as shown in Figure 8.12.

Note that when supporting functions are added to a FAST Tree, they are tiered from right to left so that the caret (^) symbols have no obstructions above them except those that have a number after the caret symbol.

PREVENT FILAMENT OXIDATION

Step 10. Develop Secondary Path Functions

Copy your latest logic diagram and paste it at the bottom of your working file, and use the Why-How Logic to develop your secondary path functions. Use the function clusters to aid you in doing this. As functions within function clusters are merged, post a "10" near the left margin to indicate that they were merged during Step 10. If perchance any clusters remain that have not been merged, copy and paste them at the end of the working file. Then expand these remaining clusters by asking "why" of the highest level function. Each time you add a new why function, ask "why" of it until you are able to discover where these clusters fit into the diagram.

Step 10 for Project 1. Develop Secondary Path Functions

A copy of the logic diagram in Figure 8.12 is copied and pasted at the end of this file and then identified as Figure 8.13. After applying the Why-How Logic to all of the supporting functions while at the same time considering all the remaining function clusters, the logic diagram is expanded until it looks like

```
Provide Luminous Energy
        Produce Light
                Convert Energy
                |   Heat Filament
                        Supply Power
                                Conduct Current
                                        Connect Lead-in Wires
                                                Mount Lamp Electrically
                                                        Twist Lightbulb
                                        ^Insulate Conductors
                                                Install Insulated Wires
                                ^Produce Electricity
                                        Rotate Armatures
                |   ^1 Position Filament
                        Position Support Wires
                                Position Button
                        ^Mount Lamp Mechanically
                                Twist Lightbulb
                |   ^2 Reduce Base Temperature
                        Deflect Heat
                                Install Reflector
                |   ^3 Prevent Filament Oxidation
                        Exclude Oxygen
                                Constrain Gases (dup)
                                        Enclose Components
                                                Add Transparent Container
                                                        Furnish Glass Bulb
                                        Provide Airtight Seal (dup)
                                                Mold Base to Bulb (dup)
                |   ^4 Prevent Filament Evaporation
                        Insert Inert Gas
                                ^Constrain Gases (dup)
                                        Provide Airtight Seal (dup)
                                                Mold Base to Bulb (dup)
```

Legend:
Vertical line identifies a given tab space to aid reading.
Caret (^) symbol indicates a supporting function.
A supporting function supports function above ^ symbol.
Numbers 1 to 4 all support the same function.

Figure 8.13.   Lightbulb FAST Tree 4

FAST Tree 4 shown in Figure 8.13. Note that each function within a cluster that is merged into the FAST Tree during Step 10 is updated by placing a "10" near the left margin. Also note that a supporting function may also have a function that supports it.

Since all the functions within the function clusters developed during Step 8 have been included in Figure 8.13 except the second cluster, it is copied and pasted during this step so it can be investigated, as shown below. Since it has not yet been merged into the logic diagram, we have to apply the Why Logic Question to its highest level function. Each time we obtain an answer, we evaluate the results to see if we can determine where it fits. We repeat this process until we are able to determine what to do with it.

~~Make Life More Enjoyable~~
~~Meet Consumer Expectations~~
~~Make More Products~~
~~Increase Productivity~~
~~Remove Obstacles~~

You may recall that we started with the basic function "make life more enjoyable" and determined that this function was not within the scope of our project. The above analysis informs us that this function cluster is out of scope and can be crossed out once we come to this conclusion, as shown above.

We can conclude, after following these 10 steps, that we have developed a somewhat realistic logic diagram of an incandescent lightbulb. The purpose, as stated at the beginning of this sample project, was to teach you how to create logic diagrams and to demonstrate the steps required to complete the task. One final comment about FAST Trees: each one is different, and often a legend should be included so others can understand any symbols. A legend was added to FAST Tree 4 for the lightbulb project in Figure 8.13 to aid in its interpretation.

In 1965, I presented my first FAST Diagram. It was a diagram of this same incandescent lightbulb. An updated diagram is shown in Figure 8.14. FAST Tree 4 shown in Figure 8.13 includes more secondary or supporting functions, but the diagrams are basically the same.

The procedure I used back in 1965 did not concentrate on creativity. It was devoted primarily to the logic required to organize functions into a logic diagram. The title of my first presentation was "Basic Function Determination Technique."[1] This technique was the inspiration that allowed me to develop my first FAST Diagram. The higher level function questions and other creative innovations were developed as I continued to train people to look at functions instead of parts.

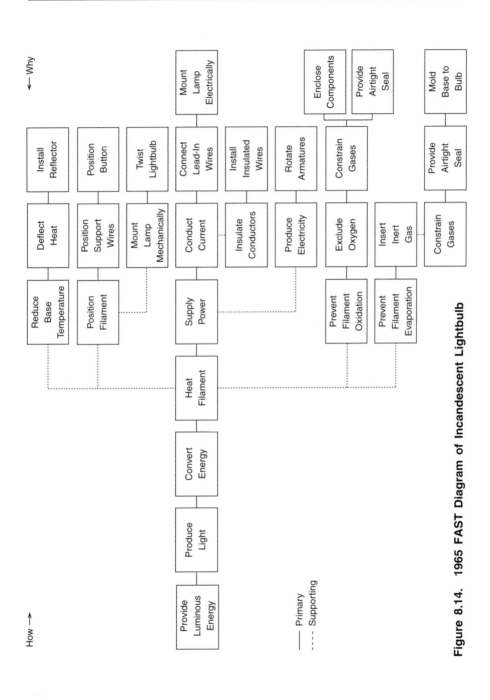

Figure 8.14. 1965 FAST Diagram of Incandescent Lightbulb

After all functions have been included, make sure you check the entire diagram to see that the logic agrees in both directions. If it does not agree, make any modifications that seem appropriate.

As function analysis and FAST diagramming have been adopted as an excellent way of performing an analysis, it has been determined that greater insight into projects can be obtained when the verb "provide" is avoided. For example, "provide transportation" can be changed to "transport people" or "transport cargo." These functions are much more informative. Also, some people feel that actions performed by people or things should not be recorded as functions. They recommend concentrating on the object of performing those actions and formulating new functions for them instead. To do so requires considerably more deep thinking on the part of the analyst and requires greater skill than I intend to present in this book.

USE SEARCHLIGHTS AND REFLECT LIGHT BEAM

Step 11. Brainstorming Higher Level Functions

Once the logic diagram has been completed, you have the opportunity to take a creative look at all the functions listed in the diagram. Generally, this step is only performed on the higher level functions of the primary path functions. Begin by inserting the highest level function into the following How Else Question:

How else can *this function* be accomplished or performed?

As you answer this question, you divorce yourself from the project and list every idea that might perform the function you insert into this question. This activity is called brainstorming the function. Brainstorming was developed by Dr. Alex Osborn in the early 1950s. It is very effective when applied to a function because a function tells the analyst what needs to be accomplished without identifying the method of accomplishment. A team of four or five people is generally selected to participate in this activity. The ground rules and concepts for performing brainstorming are recorded in Figure 8.15 and should be reviewed before starting a brainstorming session.

Once a list of ideas and concepts has been creatively developed and documented, the next task is to creatively modify each item in the list, if at all possible, so it can successfully perform the function you are investigating. Those that look promising are developed further and then evaluated to see if they can be used to develop a better concept than the concept conveyed by the

1. Every idea or suggestion is to be recorded.
2. No criticizing of anyone else's ideas or suggestion is allowed. As a reminder of this rule, I like to have everyone write "I will not criticize!" on a full sheet of paper and then crumple the paper and put it in their pocket or keep it in one hand while the brainstorming session is taking place.
3. Divorce yourself from the present project.
4. Play different roles as you think of the function.
5. Ignore what you have experienced or learned in school.
6. Present silly or ridiculous ways of doing things.
7. Disregard standards and traditions.
8. Hitchhiking on other ideas is allowed.
9. Improvements to ideas presented are encouraged.
10. Omit anyone from your group who might be intimidating, if possible.
11. Consider how physical and life sciences would perform the function.
12. Consider primitive and mass-production methods.

Figure 8.15. Ground Rules for Brainstorming

completed logic diagram. In some cases, it may be desirable to develop an entirely new logic diagram to explore a new concept. Other times, alternate methods of performing a given function within an existing logic diagram may be considered or added to the diagram. The example projects in Chapters 9 through 12 demonstrate how logic diagrams are modified to reflect other methods of performing certain functions.

Several years ago, while six teams were being trained, I was asked to help a team that had performed a brainstorming session on a function. During the brainstorming session, the team generated a list of 20 or 30 ideas by concentrating on the function "make connection." The project was a wiring harness for a missile system. I looked down the list of ideas, and the word "zipper" jumped out at me. I immediately drew a picture of a zipper made of copper or brass links, like the ones I had seen on my winter coats. Then I said, "Let's modify this now so it will work as a connector!" I made every other link of each half of the zipper a different color and said, "What if we made these dark-colored links out of copper so they can conduct electricity and these light-colored links out of plastic so they act as insulators?" Suddenly the light turned on for all the team members, and just like that they became enthusiastic about their project. You always want to try to modify every idea the team lists so it can perform the function you are brainstorming. After the seminar was over, several of the members of the team attempted to have this zipper connector patented.

After you have explored all the concepts obtained by brainstorming, you might want to consider two additional creative steps. The first of these is to see if any of the functions can be generalized, thereby removing any restrictive or limiting factors expressed by the name given to a function. The second step shows how to write functions without describing the method recorded on a completed logic diagram. These two steps are covered in Chapter 13.

Step 11 for Project 1. Brainstorming Higher Level Functions

The first question we always ask is about our basic function:

Q   How else can ***provide luminous energy*** be accomplished or performed?

Several ideas pop into my mind as I think about this question as I conclude this chapter. Once again, I will role-play in order to obtain these answers or ideas as they are conceived within my mind. The role I play and the answers I obtain while answering this question are as follows:

Role 1	Day-shift foreman
A1	Let's put skylights in the roof of the building.
A2	Let's put windows on all outside walls.
A3	Let's put mirrors that will reflect the sun's rays into the building.
A4	Let's use sensors at the top and bottom of the windows so mirrors can be positioned continuously throughout the sunlight hours and reflect the sun's rays through the windows.
A5	Let's locate mirrors at ground level and roof level on the sides of the building so the sun's rays can be utilized throughout the entire day without the building obstructing the sun's rays. Sensors can be used to aid in the proper positioning of the mirrors.
Role 2	Night-shift foreman
A1	Let's use searchlights to provide luminous energy to buildings.
A2	Let's use one searchlight for each floor of a building by shining its beam parallel to the ceiling and parallel to an

outside wall. Then let's use mirrors to reflect portions of the light beam 90 degrees at various locations to diffuse the light throughout the entire floor.

A3   Individual workstations may improve their lighting by installing mirrors tilted so as to capture the reflected rays and send those rays through light diffusers.

Role 3   Person lost in the woods or wilderness at night

A1   Turn on a searchlight and beam it into the sky from a location where help is available so a person who is lost can easily see it from a mile or so away.

A2   Beam two searchlights into the sky so a lost person has a line of reference to move toward in order to reach help or safety. This would provide the lost person a line of reference instead of a single point.

A3   Provide this same line of reference during daylight hours using sound waves by having vehicles equipped with sirens drive back and forth continuously between two points in the vicinity where the lost person was last known to have been.

Many more ideas could be generated by a team of participants, but this little exercise gives you some idea of what can be conceived within a short amount of time when a function is brainstormed. Many people limit their thinking when they perform the brainstorming exercise. I let it take me anywhere it wants to take me. Notice that when I switch roles, I hitchhike on any earlier ideas and continually try to improve on any ideas already mentioned. Try it. You'll like it.

NOTE

1. Bytheway, C.W., "Basic Function Determination Technique," SAVE Proceedings, Fifth National Conference, Vol. II, 1965, pp. 21–23.

9

PROJECT 2: TIMING DEVICE*

This chapter involves a research and development hardware project for a very complicated mechanism. It is presented so the reader will have an appreciation for what can be accomplished by persistently role-playing over and over again as the basic concepts of FAST are applied. The end result of this analysis was the replacement of 16 parts by a single part.

USING METHOD 1

Step 1. Selecting a Project

This timing device was selected initially as a project for Sperry Univac's value engineering department because our defense division was awarded the contract to develop a timer for the Briteye Flare for the United States Navy during the Vietnam War. A breadboard unit using push buttons for setting the time (as shown in Figure 9.1) had already been constructed by our facility. I was told that the linkage that allowed it to function was taken from a design used by the Navy on torpedoes during World War II. It was concluded that the breadboard unit was too complicated and had too many parts because every time setting required its own push button. Therefore, the design was changed to a round dial

---

* C.W. Bytheway, "Simplifying Complex Mechanisms During Research and Development," SAVE Proceedings 1968 International Conference, 1968, pp. 233–242.

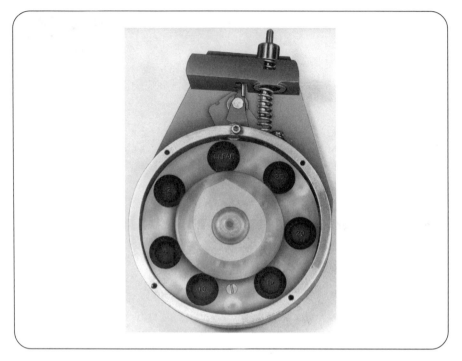

Figure 9.1. Breadboard Timer

whereby the time could be set at any time duration between a minimum and maximum time span. However, the release mechanism remained the same.

The clock escapement mechanism is secured within the round cavity of the housing, as shown at the bottom of Figure 9.2. Note that there is an arming lever over a plunger in the center of the clock mechanism. When the arming lever moves to the left, this plunger springs out and the clock starts, causing the timing disc to rotate.

The timing lever rides on the outer periphery of the timing disc until it reaches the hole in the timing disc, as shown in Figure 9.3. This picture also shows the arrangement of the other internal parts within the housing. At the instant the spring-loaded timing lever pivots into the hole of the timing disc, it causes the sear lever to be released. This in turn causes the cocking shaft to rotate because of the off-center spring force created by the firing pin spring. As soon as the cocking shaft rotates, it releases the firing pin, which then strikes the primer, which causes the flare to be ignited. Since the unit was still rather complicated, I was asked to work with the research engineer, Cyril F. Anderson,

Figure 9.2.   Back Side of First Prototype

to see if value engineering principles could simplify the design. We used my FAST technique to identify the functions and organize our thinking.

TWO TEAM MEMBERS

Step 2. Selecting Participants

The project engineer and I were the only two members of this team.

FIRST PROTOTYPE

Step 3. Formulating Initial Functions

Our analysis was limited to the parts mounted within the housing. Because the clock assembly or escapement mechanism was supplied by another manufac-

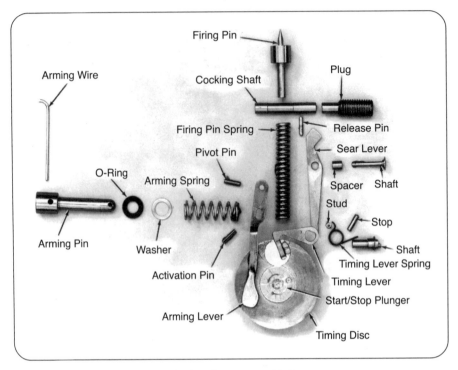

Figure 9.3. Internal Parts of First Prototype Unit

turing firm, we listed it but did not attempt to analyze the clock mechanism. Figure 9.4 lists the names of the parts within this device, along with the initial functions these parts perform.

DETONATE PRIMER

Step 4. Determine Initial Basic Function

Even though it is obvious that "detonate primer" is the basic function of this unit, we still insert this function into the Basic Function Determination Question, which yields:

> Q   If we didn't have to *detonate primer,* would we still have to perform the other functions listed?
>
> A   No

Therefore, the *initial basic function* for Project 2 is *"detonate primer."*

Parts	Initial Functions
Arming Wire	Release Arming Pin
Arming Pin	Pivot Arming Lever
O-ring	Seal Arming Pin
Arming Lever	Release Start/Stop Plunger
Start/Stop Plunger	Start Clock
Timing Disc	Release Timing Lever
Clock	Rotate Timing Disc
Timing Lever	Rotate Timing Lever Shaft
Timing Lever Shaft	Release Sear Lever
Sear Lever	Release Cocking Shaft
Cocking Shaft	Release Firing Pin
Firing Pin	Detonate Primer

Figure 9.4. Initial Functions for Timing Device

LIGHT FLARE

Step 5. Develop Higher Level Functions

The next task is to go in search of higher level functions by inserting the initial basic function into the Higher Level Function Questions, which then read as follows:

1. Why is it necessary to ***detonate primer***?

2. What higher level function caused ***detonate primer*** to come into being?

3. What is really trying to be accomplished when ***detonate primer*** is performed?

All three of these questions can be answered by the function "light flare." Usually a single answer is less than is normally expected to express the answers to these three questions, but in this case we were limited. If we could have included participants from the Navy Department, we might have obtained higher level functions but probably would not be able to disclose them because of national security. The thing a study group has to realize when it has a project like this is that it often has to limit its effort to the product it is contracted to design and build. Therefore, we limited our study.

DETONATE EXPLOSIVE

Step 6. Identify the Basic Function

Since no more functions have been added, our initial basic function is also our *basic function,* which is *"detonate primer"* or "detonate explosive."

DEVELOP HOW TO DETONATE PRIMER

Step 7. Develop Primary Path Functions

This step is started by listing all the functions formulated for this project up to this point. They are listed in Figure 9.5.

Next post the basic function and answer the following question:

Q   How is *detonate primer* actually accomplished?

Then ask the Why-How Logic Questions of any new functions and also express those answers as functions. Repeat this process to generate and organize the primary path functions that describe how the basic function is accomplished as presently designed. This is exactly what we did to develop the primary path functions.

When we asked the Why-How Logic Questions and checked each new function as it was added to the logic tree in the "why" and "how" directions, we effectively developed the primary path functions shown in Figure 9.6.

7	Release Arming Pin
7	Pivot Arming Lever
	Seal Arming Pin
7	Release Start/Stop Plunger
7	Start Clock
7	Release Timing Lever
7	Rotate Timing Disc
7	Rotate Timing Lever Shaft
7	Release Sear Lever
7	Release Cocking Shaft
7	Release Firing Pin
7	*Detonate Primer

Figure 9.5. List of Functions

```
Detonate Primer
      Release Firing Pin
            Release Cocking Shaft
                  Release Sear Lever
                        Rotate Timing Lever Shaft        ↖
  How                         Release Timing Lever          Why
      ↘                     Rotate Timing Disc
                              Start Clock
                                    Release Start/Stop Plunger
                                    Pivot Arming Lever
                                          Release Arming Pin
                                                Remove Arming Wire
```

Figure 9.6. Timing Device FAST Tree 1

PREVENT MECHANISM CONTAMINATION

Step 8. Evaluate Remaining Formulated Functions

The first thing we do in this step is to list all the functions in Figure 9.5 which have not yet been considered or included within our FAST Tree; the only one is "seal arming pin." We then highlight this function and apply the Why-How Logic Questions to it; the results of that activity are

9	Prevent Mechanism Contamination	
10	**Seal Arming Pin**	
10		Install O-ring

When checked, the logic for this cluster is found to be correct in both directions. Since only one function cluster exists, we first try to merge it into the primary path functions of FAST Tree 1. Since it cannot be merged, we move on to Step 9.

COCK FIRING PIN

Step 9. Using When/If Logic to Add Supporting Functions

A supporting function is a function that must be performed when and if a primary path function cannot be performed without that function being performed at the same time or some time prior to the primary path function being performed. We start by asking the following question. As we do this, we determine if any other supporting functions exist that we have not yet recorded or considered.

> When/if *detonate primer* is performed, what other functions must be performed?

We could reason that if we are going to detonate the primer, we are going to have to "cock firing pin" since it is spring loaded. Therefore, "cock firing pin" becomes a supporting function. When we look at the next primary path function, we answer the following question:

> When/if *release firing pin* is performed, what other functions must be performed?

We obtain the same supporting function for this answer. Since the firing pin cannot be released unless the cocking shaft is cocked by compressing the firing pin spring, we attach "cock firing pin" as a supporting function of "release firing pin." Before we insert this supporting function into our FAST Tree, we copy and paste the tree at the end of the working file. We continue to ask this When/ If Question of all the other primary path functions. During this exercise, we add "wind clock" and "cock start/stop plunger" as supporting functions.

When we ask "when" and "if" of "pivot arming lever," we are able to add the supporting function "prevent mechanism contamination." This function appears in the function cluster developed during Step 8. Since this function appears in the function cluster developed during Step 8 but is not inserted until Step 9, a "9" is posted near the left margin to indicate which step added this function to the FAST Tree. Figure 9.7 shows FAST Tree 2 with these supporting functions as part of the diagram.

Note that each supporting function is added to a FAST Tree by placing a caret (^) symbol directly below the primary path function it supports instead of putting the supporting function one tab space to the right of the function it supports, as is done in the function cluster. Vertical lines are frequently added to ensure that a given tab space is correctly interpreted.

DEVELOP HOW TO COCK FIRING PIN

Step 10. Develop Secondary Path Functions

After the supporting functions have been added, we once again copy and paste the logic diagram at the end of the working file before we expand each sup-

```
Detonate Primer
  | Release Firing Pin
  |    Release Cocking Shaft
  |       Release Sear Lever
  |          Rotate Timing Lever Shaft      ↖
  |  How      Release Timing Lever        Why
  |      ↘         | Rotate Timing Disc
  |                | Start Clock
  |                |    Release Start/Stop Plunger
  |                |       Pivot Arming Lever
  |                |          Release Arming Pin
  |                |             Remove Arming Wire
  |                |             ^Prevent Mechanism Contamination
  |                |    ^Cock Start/Stop Plunger
  |                | ^Wind Clock
  | ^Cock Firing Pin
```

Figure 9.7.   Timing Device FAST Tree 2

porting function branch. They are expanded by asking the How Logic Question of all supporting functions and then asking the Why-How Questions of any new functions over and over again. Since the last two functions of the function cluster developed during Step 8 are added during this step, a "10" is posted near the left margin next to each of these two functions. Next we repeat Steps 9 and 10 until the supporting function paths have been developed. Once this is done, the diagram is not considered to be complete until the logic of each function in the entire diagram has been checked. The completed logic diagram for this timing device is displayed as FAST Tree 3 in Figure 9.8.

I have been extremely repetitive in this chapter as we performed each step so you would remember to follow the steps and always copy and paste your work before you alter it. This allows you to always have a continuous record of the logic you used to develop your diagram and other important information used to perform your analysis.

RELEASE COCKING SHAFT

Step 11. Brainstorming Higher Level Functions

This step suggests that we insert the primary path functions into the following question:

```
Detonate Primer
  Release Firing Pin
      Release Cocking Shaft
          Release Sear Lever
              Rotate Timing Lever Shaft
                  Release Timing Lever
                      Rotate Timing Disc
                        Start Clock
                            Release Start/Stop Plunger
                                Pivot Arming Lever
                                    Release Arming Pin
                                        Remove Arming Wire
                                    ^Prevent Mechanism Contamination
                                        Seal Arming Pin
                                            Install O-ring
                            ^Cock Start/Stop Plunger
                                Rotate Timing Disc
                                Position Arming Lever
                                Compress Plunger Spring
                                Mount Cover Plate
                      ^Wind Clock
  ^Cock Firing Pin
      Compress Firing Pin Spring
      Rotate Cocking Shaft
      Position Sear Lever
      Latch Timing Lever
          Position Timing Disc
              Rotate Clock Assembly
                  Rotate Clock Housing
              ^Seal Clock Housing
```

Figure 9.8. Timing Device FAST Tree 3

How else can ***this function*** be performed or accomplished?

Before we do this, it is a good idea to convert our FAST Tree into a FAST Diagram so we can visually see how all the functions fit together. Figure 9.9 illustrates how the timing device FAST Tree 3 is converted into a FAST Diagram.

One of the best ways to approach the brainstorming task is to post the primary path functions along with the parts that actually perform those functions and display them as a Partial FAST Diagram. Figure 9.10 shows these functions

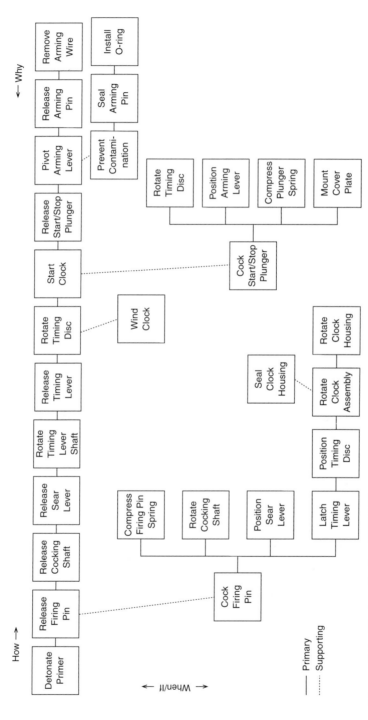

Figure 9.9.   Timing Device FAST Diagram

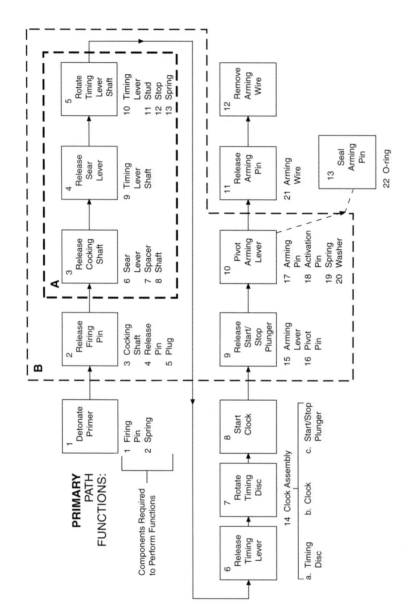

Figure 9.10.   Partial FAST Diagram

along with the corresponding parts. We included the supporting function "seal arming pin." Note that the part that actually performs each function is underlined. We used the Charles Kettering approach, where we imagine ourselves as one of the parts.

For example, as we went down the list of primary path functions, we asked the following question and then looked at the underlined parts:

Q   How else can ***release cocking shaft*** be performed or accomplished?

MODIFY TIMING LEVER

As we looked at the various parts, we tried to modify them. With a little creativity, we were able to modify the timing lever identified in Figure 9.3 so it could also perform the function "release cocking shaft." When this was accomplished, we also had to modify the arming lever. The result of this creativity appears in our second prototype design, shown in Figure 9.11.

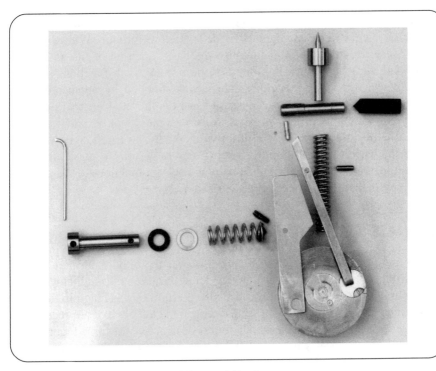

Figure 9.11. Internal Parts of Second Design

TIMING LEVER RELEASES START/STOP PLUNGER

This new timing lever actually performed the three functions enclosed within block A, as shown in Figure 9.10. This new design looked very promising, but the more we looked at the functions, we decided that we could get this new timing lever to also pivot the arming lever which releases the start/stop plunger and thereby start the clock.

As we made these changes to the timing lever, we realized that with a few more minor changes we could get the timing lever to also release the firing pin at the same time. All the functions within block B of Figure 9.10 were then accomplished by a new redesigned timing lever. We then became aware of the fact that we did not need items 18, 19, and 20 listed under function 10 in Figure 9.10 because the firing pin spring could also force the arming lever off the start/stop plunger. When all of these design changes were incorporated into this new timing lever, we were able to build our third design, shown in Figure 9.12.

Note that the removal of the arming wire permits the firing pin spring to rotate the timing lever, which enables the start/stop plunger to pop out, which starts the clock. It also forces a tang of the timing lever to ride on the periphery of the timing disc. When this tang is forced into the hole in the timing disc by the cocked firing pin spring, the timing lever rotates enough to release the firing pin, which then strikes and ignites or detonates the primer. With this new design, we were able to still accomplish the basic function of this unit.

We had no control regarding three of the primary path functions because they were performed by the clock, which was supplied by another defense contractor. At this point in our analysis, we had eliminated 15 parts as a result of analyzing functions instead of parts. Most designers look at parts to see if they can be manufactured using a different method. The FAST approach first looks at the basic function to see if a better way to perform it exists. If that is not possible, then we brainstorm the other higher level functions of the primary path to see if they can be accomplished better.

ELIMINATE ARMING PIN

The arming pin at the left side in Figure 9.12 was the only major part we had not realistically eliminated. After thinking about this, we realized that we could flip our timing lever over and allow it to also perform this arming pin function. This became possible by adding a second tang to the left edge of our redesigned timing lever, which then became our fourth design, as shown in Figure 9.13. When the arming wire is inserted behind this tang, it prevents the lever from

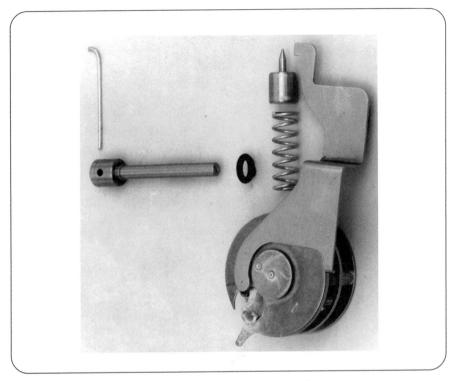

Figure 9.12.   Internal Parts of Third Timer Design

rotating until the arming wire is removed at the time the Briteye Flare timing device is ejected from the aircraft.

These changes in our third and fourth designs made it possible to simplify the housing and to eliminate the arming pin. It should be noted that the frictional force required to pivot the timing lever off the top of the start/stop plunger is much larger than the maximum force allowed against the periphery of the timing disc once the clock started to run. Therefore, the firing pin spring had to be able to produce two forces of different magnitudes within acceptable tolerances. This was accomplished in these last two designs by pivoting the timing lever at point A to start the clock. Then, as the lever moves, it contacts the housing at point B, which becomes the pivot point for the lever, as shown in Figure 9.13. This change in pivot points reduces the force of the tang against the timing disc so the frictional force is less than the torsion clock spring force rotating the disc, which allows the clock to run.

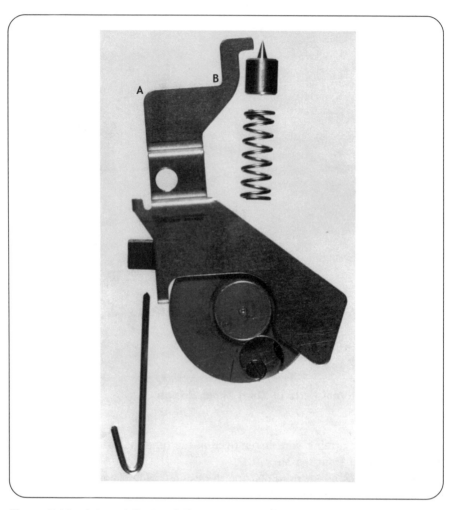

Figure 9.13. Internal Parts of Fourth Timer Design

SPRING STEEL TIMING LEVER

After you get on a creativity binge like we were, you begin to wonder what other bright ideas you might come up with. It was about that time when I thought, "Let's make the timing lever out of spring steel, and then we can eliminate the firing pin spring and perhaps the firing pin." In order for this idea to work, we

Figure 9.14. Spring Steel Timing Lever

needed two leaf springs as an integral part of any design. We finalized on the design shown in Figure 9.14. The flat horizontal spring at the far right provided enough force to start the clock. When deflected, the thinner and more flexible curved flat spring stored the energy required to accelerate the timing lever's mass so it would detonate the primer.

This flexible spring also produces the proper force for the tang that rides on the periphery of the timing disc. When this tang reaches the hole in the timing disc, it allows the stored energy in the spring to accelerate the lever's mass, which detonates the primer. It is hard to believe that this spring steel timing lever is capable of replacing the 18 parts shown in Figure 9.15. These are the initial parts we were challenged to simplify.

Figure 9.16 shows this new timing lever in its impact position within a Lexan plastic housing after the set time has elapsed. We were surprised when the supplier of these springs could only guarantee that one out of every three springs would meet our specifications because some springs warped too much

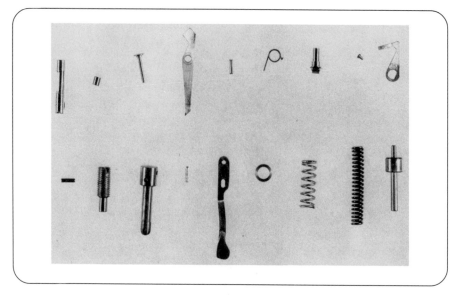

Figure 9.15.   18 Parts Replaced by Spring Steel Timing Lever

during the heat-treating process. Even though we had been very creative, we had to become realistic. You also have to be realistic whenever you are seeking a creative solution. The following chapters will show you how to do this as we analyze a variety of projects.

MUST BE PRACTICAL

When we made a comparison of the costs of our five designs, we discovered that our fourth design was our most economical. The bar chart in Figure 9.17 shows a cost-wise comparison of these five designs. Note that our functional approach still reduced the cost of the first prototype's internal parts by 80 percent.

It was obvious after doing this cost analysis that the fourth design should become our production unit. The 13th production unit is shown in Figure 9.18. This unit required only 6 internal parts compared to the 22 we were asked to try to simplify; we were not allowed to change 2 of those 6 parts. We had no control over the clock, and the arming wire which had to be attached to the aircraft was a requirement so the timer could be activated when the unit was

Figure 9.16.   Assembly of Fifth Timer Design

deployed from the aircraft. In reality, we had only 20 internal parts to work with and were able to reduce them to only 4. In other words, we eliminated 16 parts using function analysis combined with FAST diagramming.

Keep in mind that during this entire exercise we concentrated on finding other ways of performing the primary path functions shown in Figure 9.10 by role-playing. We imagined ourselves being that tiny timing lever. The basic function was "detonate primer." How did we do that? We released the firing pin. How did we release the firing pin? We pivoted or rotated the redesigned timing lever.

Steps 12 and 13 of the FAST procedure (which are covered in Chapter 13) were not required for this project. The purpose of these last two steps is to stimulate additional creativity. We figured that we had been creative enough.

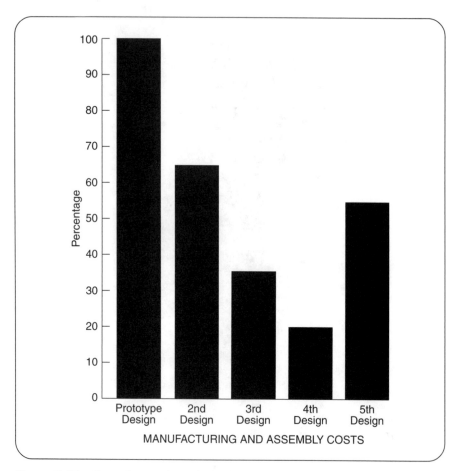

Figure 9.17. Cost Comparison Chart

I have intentionally gone into a lot of detail as I have described the various creative developments we made as we went from one design to the next. You would probably have to be an engineer to do a similar mechanical project. However, I believe anyone can be as creative as we were if they will select a function and then think of various roles to play as they ask the How Else Questions about that function. Various examples of this approach are presented in Chapters 10 through 13.

Figure 9.18.   Production Unit

10

PROJECT 3: LOVE

This chapter uses a very different example than the previous chapters. It ventures into human relationships of feelings toward each other rather than examples of hardware or procedures. It demonstrates the versatility of FAST to take on practically any subject that can be conceived. The example used in this chapter could have been among friends, co-workers, teammates, or any similar relationship. As a learning exercise, the example used here is a fictitious Western family where family members have equal rights with each other. It is recognized that some cultural customs may not identify with a family as thus presented; however, the principles of developing a FAST Diagram for a very different subject can be learned from this example.

We begin by taking two short sentences about respect and love for each other and developing a logic diagram that expands to 23 functions toward fostering better relationships among people. Twenty-eight other functions were formulated by brainstorming just one function in the primary path. In addition, Step 13 of the FAST procedure adds an additional 13 functions that bring the total to 64 functions. Many more could be formulated, but this is sufficient to demonstrate how a few facts can quickly be creatively expanded by applying the FAST Creativity technique. This chapter also presents a discussion on how to utilize a thesaurus as an aid in naming functions.

USING METHOD 2

Step 1. Selecting a Project

Method 2 is used to select Project 3. Method 2 is as follows:

Write one, two, or three sentences about the project. The information contained within these sentences is used to formulate the project.

The following two-sentence quote will be used to establish the project in this chapter:

The opposite of love is selfishness. The opposite of love is trying to control another person.

—*Arthur E. Millican*

ONLY ONE PARTICIPANT

Step 2. Select Team Members

We will assume that only the author will analyze this particular project. However, I will use the pronoun "we" throughout this project, as if several people were participating.

EXPRESS LOVE

Step 3. Initial Functions

This step takes the information contained within Step 1 and converts it into functions. In Chapter 2, we converted the first sentence of this project into the following two functions:

Express Love
Avoid Selfishness

Since both sentences of this project talk about love and love suggests that peace and harmony must exist if love is to abound, I have intuitively selected a role to play in order to name the initial functions. That role is a person who is trying to promote love between spouses.

The last sentence of this project suggests that controlling be avoided, which yields:

Avoid Controlling Someone

Our three initial functions for Project 3, therefore, become

Express Love
Avoid Selfishness
Avoid Controlling Someone

As a person approaches any project, he or she intuitively selects a role to play. The role a person selects may be himself or herself or a companion, boss, neighbor, or best friend. It might be a doctor, a lawyer, or an engineer. A person may even switch roles as he or she tries to arrive at worthwhile initial functions. Since role-playing was covered in considerable detail in Chapter 7, it is sufficient to state that every time you name a function, you intuitively select a role.

Thus far, we have identified only three initial functions for Project 3: *"express love," "avoid selfishness,"* and *"avoid controlling someone."*

USING THE COMPUTER THESAURUS AND DICTIONARY

At this point in our search for initial functions, the use of the thesaurus and dictionary is introduced. These are very worthwhile tools because they aid you in selecting the best possible verb and noun for each function. Once you have identified the initial functions for a project, it is a good idea to use the thesaurus to check the verbs you have recorded. When we look up the verb "express," which is the verb used to name the function "express love," we find about eight verbs that could be used as synonyms and three general definitions, which produce more synonyms.

The first definition given for the verb "express" is to give expression to, as by gesture, facial aspects, or bodily posture. The thesaurus informs us that we may replace express with display, convey, manifest, or communicate.

The second definition for the verb "express" means to put into words; therefore, we may replace express with state, say, tell, voice, vent, utter, convey, declare, articulate, communicate, enunciate, or vocalize.

The third definition for the verb "express" means to extract liquid from by applying pressure, to utter publicly, or to convey in language or words of a particular form.

After investigating the three definitions for the verb "express" using the thesaurus and the dictionary, we conclude that the initial function "express love" should be changed to "convey love" since "convey" is the synonym that includes all three of these definitions. Also, it encompasses much more than just verbally expressing one's love. The verb "avoid" appears to be an excellent verb for the role I am playing since the thesaurus only list nine synonyms, none of

*Convey Love
Avoid Selfishness
Avoid Controlling Someone

Figure 10.1. Initial Functions for Love Project

which seem to formulate into a better function. Therefore, the final three initial functions we will use for this project are shown in Figure 10.1.

Some participants who participate with you may select entirely different verbs as well as nouns for this project. If they do, then a discussion should take place until all agree on the initial functions. It will become obvious if participants are playing different roles.

Generally, it is not necessary to use the thesaurus and dictionary and go through this two-step operation when selecting the best verb, although they are readily available in most word-processing programs just by clicking the mouse. Someone who is unsure which word would be the best selection can access the thesaurus and dictionary before making the final selection.

CONVEY LOVE

Step 4. Initial Basic Function

Once the initial functions have been named, the next task is to determine which of these functions is the initial basic function. The greatest benefits come when you devote your energies and creative abilities to the basic function, which is usually located at a higher level above the initial basic function. For that reason, we want to find the initial basic function from our list of three functions so we can use it as the springboard to higher level logic functions. Therefore, we now apply the Basic Function Determination Technique to our initial functions.

"Convey love" appears to be the function we should select as the initial basic function; therefore, it is inserted into the Basic Function Determination Question and each of the other functions is evaluated with respect to it. In essence, we are mentally thinking:

Q1   If *convey love* didn't have to be performed, would *avoid selfishness* still have to be performed?

A1   No

Q2   If **convey love** didn't have to be performed, would **avoid controlling someone** still have to be performed?

A2   No

Because both of these questions yield a "no" answer, "*convey love*" becomes the *initial basic function* and the functions "avoid selfishness" and "avoid controlling someone" automatically become dependent functions or lower level functions. An asterisk (*) is placed next to "convey love" in Figure 10.1 once this decision has been made.

FOSTER BETTER RELATIONSHIPS

Step 5. Develop Higher Level Functions

We are normally looking for at least three new facts when we perform this step. One also has to be realistic when looking for higher level functions. I still remember my boss getting upset with me because I wanted to know why we were building a certain missile for the United States Army. He pointed out to me that almost everyone's job in our division was being funded by this missile system. Why would we want to eliminate our own jobs? We may not want to document our answers, but the following three questions are still good questions to ask. Note that as we ask these three higher level logic questions, we switch roles until they yield good answers.

Q1   Why is it necessary to **convey love**?

Role A person with a spouse and five children
A1   I want to increase my relationship with those close to me

Q2   What higher level function caused **convey love** to come into being?

Role A parent with a wayward child
A2   There is too much tension among family members

Q3   What is really trying to be accomplished when **convey love** is performed?

Role Someone at work working with other people
A3   I'm trying to treat other people as I would like to be treated

This step is not complete until you have changed these answers into functions. The functions I formulated from these three answers were "foster better relationships," "remove tension," and "treat people respectfully."

USING THE BASIC FUNCTION DETERMINATION TECHNIQUE

Step 6. Identify the Basic Function

The first task in this step is to list the initial basic function and add to it the new functions obtained during Step 5, as shown in Figure 10.2. Then select from the list the function that appears to be the most important or basic. Use it in the Basic Function Determination Question, and if all answers are no, then mark your selection with an asterisk (*).

> Q   If we didn't have to *foster better relationships*, would we still have
>   to *convey love, remove tension,* or *treat people respectfully*?
> A   No

Therefore, "*foster better relationships*" becomes the new basic function. Obviously, you don't have to write down the question. All you have to do is mentally insert the selected function into the question and look at the other functions. If all answers are no, you have selected the right function.

DEVELOP HOW TO FOSTER BETTER RELATIONSHIPS

Step 7. Develop Primary Path Functions

Before starting the logic diagram, it is a good idea to list all the functions you have formulated thus far for your project and identify your basic function in the list, as shown in Figure 10.3.

Convey Love
*Foster Better Relationships
Remove Tension
Treat People Respectfully

Figure 10.2.   Higher Level Functions

```
              Convey Love
              Avoid Selfishness
              Avoid Controlling Someone
        7     *Foster Better Relationships
        7     Remove Tension
        7     Treat People Respectfully
```

Figure 10.3. Formulated Functions

Next, start to develop the logic diagram by posting the basic function as the first entry in your FAST Tree; ask the How Question of it, and then ask the Why-How Logic Questions of each new function formulated. Continue this process until you are unable to think of any new functions. Since we are not trying to solve the world's problems with this project, we have elected to limit the number of functions listed. We are more interested in showing how various aspects of a given subject can be expanded very quickly by using the Why-How Logic than we are in developing all branches of the tree. The primary path functions we developed during this step are shown in Figure 10.4

As primary path functions are added to the FAST Tree, we type a "7" near the left margin in Figure 10.3 to show which functions have been included during this step. Also, every time we add a new function by asking the How Logic Question, we check to see if it agrees with the Why Logic Question. As we consider the functions in Figure 10.4, we should ask the following questions:

```
          Foster Better Relationships
            Treat People Respectfully
              Display Courtesy
              Speak Kindly
              Use Words of Encouragement
            Remove Tension
              Be More Understanding
              Speak Reverently
              Listen Compassionately
              Offer Helping Hand
```

Figure 10.4. Love FAST Tree 1

Q1   Why *display courtesy*?
A1   Treat people respectfully

Q2   Why *speak kindly*?
A2   Treat people respectfully

Q3   Why *use words of encouragement*?
A3   Motivate people

Q4   Why *motivate people*?
A4   Foster better relationships

Every function in the why direction needs to be checked. Occasionally this exercise produces a new function, as demonstrated in Question 3. When this happens, we modify the logic diagram so that every function agrees with our logic analysis. Our modified diagram is shown in Figure 10.5.

Keep in mind that we are always trying to expand our understanding of the subject. When a why answer already exists, it is easy to agree with the answer that is posted one tab space to the left and above it. Try to think more deeply and role-play as you answer the Why Logic Question. Also, check the logic by using the Verification Question. Any one of the tree branches in Figure 10.5 could be expanded by asking the How Logic Question of the lower level functions. However, our purpose here is to learn how to develop a logic diagram; therefore, we will move on to Step 8.

Foster Better Relationships
Treat People Respectfully
Display Courtesy
Speak Kindly
Motivate People
Use Words of Encouragement
Remove Tension
Be More Understanding
Speak Reverently
Listen Compassionately
Offer Helping Hand

Figure 10.5. Love FAST Tree 2

TREAT PEOPLE RESPECTFULLY

Step 8. Evaluate Remaining Formulated Functions

Three functions in Figure 10.3 have not been included in FAST Tree 2; there-fore, we apply the Why-How Logic to each one of them to develop clusters in order to find out where they fit in the diagram, as demonstrated below:

8	Treat People Respectfully
8	**Convey Love**
8	Avoid Controlling Someone
8	Avoid Selfishness

8	Convey Love
8	**Avoid Selfishness**
8	Share Financial Matters

8	Convey Love
8	**Avoid Controlling Someone**
8	Honor Free Agency

The next task is to check the logic to see if any of the how functions are supporting functions by asking the following questions:

Q   Does *convey love* help *treat people respectfully*?
A   Yes

Q   Does *avoid controlling someone* help *convey love*?
A   Yes

As we check the rest of these function clusters, we discover that there are not any supporting how functions within them. Since none exist, we then see if any of these clusters can be merged into the primary path. Inasmuch as the first cluster includes one of the primary path functions, we copy and paste the FAST Tree at the end of the file and merge this cluster, as shown in Figure 10.6. As this and other clusters are also merged during this step, an "8" is placed near the left margin anywhere the clusters were developed.

After the function clusters have been merged, it is a good idea to check the logic of all new functions added to the diagram before proceeding to the next step.

```
                    Foster Better Relationships
                      Treat People Respectfully
                          Display Courtesy
                          Speak Kindly
                          Convey Love
                              Avoid Controlling Someone
                                  Honor Free Agency
                              Avoid Selfishness
                                  Share Financial Matters
                      Motivate People
                          Use Words of Encouragement
                      Remove Tension
                          Be More Understanding
                          Speak Reverently
                          Listen Compassionately
                          Offer Helping Hand
```

Figure 10.6. Love FAST Tree 3

BE FRIENDLY AND SPEAK SOFTLY

Step 9. Using When/If Logic to Add Supporting Functions

We apply this step to every function listed in Figure 10.6, since every function is in the primary path of the basic function. In this case, the primary path has several branches. We start by asking the following questions:

Q1 When/if *foster better relationships* is performed, what other functions must be performed?

A1 None come to mind

Q2 When/if *treat people respectfully* is performed, what other functions must be performed?

A2 None come to mind

Q3 When/if *motivate people* is performed, what other functions must be performed?

A3 Be friendly

As we continued this process, we added three supporting functions, as shown in Figure 10.7.

Foster Better Relationships
Treat People Respectfully
Display Courtesy
Speak Kindly
Convey Love
Avoid Controlling Someone
Honor Free Agency
Avoid Selfishness
Share Financial Matters
^Be Honest
Motivate People
Use Words of Encouragement
^Be Friendly
Remove Tension
Be More Understanding
Speak Reverently
Listen Compassionately
Offer Helping Hand
^Speak Softly

Figure 10.7. Love FAST Tree 4

DEVELOP HOW TO BE FRIENDLY

Step 10. Develop Secondary Path Functions

The secondary path functions are developed the same way as the primary path functions. We ask the How Logic Question of the three supporting functions we added during Step 9 and then ask the Why-How Logic Questions of all new function we add to the diagram, as shown in Figure 10.8.

More functions no doubt could be added to this logic diagram, but we will limit the functions to those shown in Figure 10.8 for the present. We will now convert this FAST Tree into a FAST Diagram. Whenever you want to present your organized thinking to a group, it is a good idea to construct a FAST Diagram so people can visually see how all the functions fit together. The conversion of FAST Tree 5 into a FAST Diagram is shown in Figure 10.9. Appendix A lists several Visio® templates available to aid you in creating FAST Diagrams.

A FAST Diagram allows you to better see which functions you may wish to consider when you move to the next step in this procedure. Your first inclination may be that we probably already have more than enough ideas.

```
                 Foster Better Relationships
                  Treat People Respectfully
                       Display Courtesy
                       Speak Kindly
                       Convey Love
                           Avoid Controlling Someone
                               Honor Free Agency
                           Avoid Selfishness
                               Share Financial Matters
                               ^Be Honest
                                   Disclose All Transactions
                  Motivate People
                      Use Words of Encouragement
                  ^Be Friendly
                      Smile Appropriately
                      Greet Everyone
                  Remove Tension
                      Be More Understanding
                      Speak Reverently
                      Listen Compassionately
                      Offer Helping Hand
                  ^Speak Softly
                      Control Anger
```

Figure 10.8.   Love FAST Tree 5

However, you might be surprised by how many more ideas you can come up with when you combine brainstorming with role-playing. You might be thinking, "How many more ideas do we need to effectively cover this subject?" After you see what can be done, you might think otherwise.

BE PROMPT AND PROTECT CHILD

Step 11. Brainstorming Higher Level Functions

Whenever we apply brainstorming to the higher level functions of a nonhardware-type project, we generally add additional branches to the primary path. For example, if we brainstorm the function "convey love," we would ask the following question:

Q   How else can *convey love* be accomplished or performed?

How →   ← Why

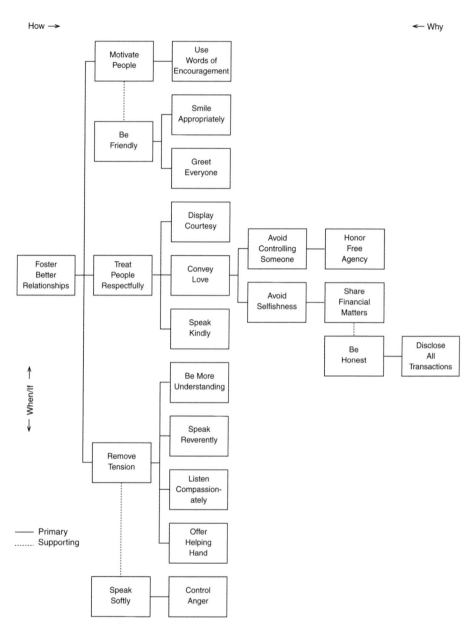

Figure 10.9. Love FAST Diagram

Role   Husband toward his wife
A1   Be prompt
A2   Be attentive
A3   Be considerate
A4   Be forgiving
A5   Be thankful
A6   Be faithful
A7   Be affectionate
A8   Appreciate cooking
A9   Kiss wife
A10   Hug wife

Role   Mother toward her child
A11   Protect child
A12   Discipline child
A13   Encourage child
A14   Praise child
A15   Entertain child
A16   Work with child
A17   Assist child
A18   Comfort child

Role   Wife toward her mother-in-law
A19   Appreciate advice
A20   Compliment appearance
A21   Telephone mother-in-law
A22   Render compassionate help

Role   Man toward a best friend
A23   Share experiences
A24   Hang out together
A25   Fish together
A26   Hunt together
A27   Enjoy sports together
A28   Express appreciation

We could play many different roles as we attempt to increase the number of ideas we generate for the function "convey love" during a brainstorming session. We could do the same thing for all the other functions in the primary path. Our purpose here is to realize how easy it is to expand one's thinking by role-playing while brainstorming functions. No attempt will be made to merge any of these 28 functions into FAST Tree 5 shown in Figure 10.8. All of these functions are of little value unless we look over the list and ask ourselves "Why should I perform any of these functions?" The majority of them, if implemented, will make life more enjoyable and make the world a better place in which to live.

THE INVENTIVE GENIUS WITHIN YOU

This sample project demonstrates how two simple sentences can be expanded into many thought-provoking areas by following the FAST procedure for developing a FAST Tree. Fifty-one functions have been formulated thus far. Many more could be formulated if we took the time to expand the logic diagram or do additional brainstorming. We have not applied Steps 12 and 13 to this subject as yet, but they will be explained and applied in Chapter 13. This expanding of information by answering the Why-How Logic Questions requires creativity. The FAST procedure provides you with that opportunity because each time you formulate a new function, you can ask "why" and "how" of that new function. You never know when a new function will spark that inventive genius within you. The functions you creatively explore always focus your mind on your project.

11

PROJECT 4:
THREE-TON HEAT PUMP*

This chapter involves a project assigned to a team for the purpose of reducing cost and making improvements to a consumer-type product used in homes and businesses.

USING METHOD 3

Step 1. Selecting a Project

Whenever you analyze a product or a piece of equipment which has already been built and is normally sold to a customer, client, or consumer, you should use Method 3 to define your project. Since the project selected for this chapter fits this definition, we proceed as follows.

The project is a three-ton heat pump. The six questions and corresponding answers outlined for Method 3 form the nucleus for starting this project:

Q1   What product or piece of equipment have you been assigned to analyze?

A1   Analyze a company's three-ton heat pump, which has the capability to heat, cool, and dehumidify a living space

---

* Initial functions and typical team members used for this project were gleaned from material supplied by Theodore C. Fowler of Dayton, Ohio.

Q2 What is the main purpose for which this product has been built or assembled?

A2 Make people comfortable

Q3 How can this product be made so it is more dependable?

A3 Make sure it doesn't deteriorate or have to be repaired very often

Q4 How can this product be made so it is more convenient to use?

A4 Make the mounting simple and the controls readily accessible

Q5 How can this product be improved?

A5 Make it more efficient

Q6 How can this product be made so it is more pleasing to the five senses?

A6 Make it more attractive and make it so a normal-size person can operate it

FIVE TEAM MEMBERS

Step 2. Selecting Participants

It was decided that a five-member team would analyze this project. The team members assigned to this project came from the departments shown in Figure 11.1.

HEAT AIR

Step 3. Initial Functions

We next formulate our initial functions by changing our answers to the six questions generated during Step 1:

A1	Analyze a company's three-ton heat pump, which is presently able to heat, cool, and dehumidify a living space
Functions	Heat air, cool air, and dehumidify air

> Engineering
> Manufacturing Engineering
> Finance
> Purchasing
> Marketing

Figure 11.1. Team Members

A2	Make people comfortable
Function	Supply comfort

A3	Make sure it doesn't deteriorate or have to be repaired very often
Functions	Limit deterioration, prevent leakage, and minimize repairs

A4	Make the mounting simple and the controls readily accessible
Functions	Simplify mounting and make controls accessible

A5	Make it more efficient
Function	Increase efficiency

A6	Make it more attractive and make it so a normal-size person can operate it
Functions	Make product attractive and accommodate normalcy

SUPPLY COMFORT

Step 4. Determine the Initial Basic Function

The first thing we do is create a list all of the initial functions, as shown in Figure 11.2, and select the function we think is the basic function at this level of our investigation. Then we insert this function into the Basic Function Determination Question and if it is verified place an asterisk (*) next to it.

From the list in Figure 11.2, we select "supply comfort" as the function we think to be the basic function and ask:

Heat Air
Cool Air
Dehumidify Air
*Supply Comfort
Limit Deterioration
Minimize Repairs
Simplify Mounting
Make Controls Accessible
Increase Efficiency
Make Product Attractive
Accommodate Normalcy

Figure 11.2.   Initial Functions for Three-Ton Heat Pump

Q   If *supply comfort* didn't have to be performed, would any of the
other functions still have to be performed?
A   No

Since all the other functions yield "no" answers, the *initial basic function*
becomes *"supply comfort."*

SATISFY CUSTOMER

Step 5. Develop Higher Level Functions

The next task is to search for higher level functions. This is accomplished by
asking questions of the initial basic function recorded during Step 4 by inserting
the initial basic function into the three Higher Level Function Questions. As we
do this, we will intuitively play a number of different roles:

Q1   Why is it necessary to *supply comfort*?

Role   Salesperson
A1   Satisfy customer

Q2   What higher level function caused *supply comfort* to come
into being?

Supply Comfort
Satisfy Customer
Increase Sales
*Make Profit

Figure 11.3. Higher Level Functions

Role   Salesperson
A2   Increase sales

Q3   What is really trying to be accomplished when *supply comfort* is performed?

Role   Chief executive officer
A3   Make profit

These three answers yield three new functions. When we combine these functions with the initial basic function, we obtain the functions listed in Figure 11.3.

MAKE PROFIT

Step 6. Identify the Basic Function

Q   If *make profit* didn't have to be performed, would any of the other functions still have to be performed?

Role   President of the company that manufactures the heat pump
A1   No

Role   Purchaser of the heat pump
A2   Yes

Whenever you answer the Basic Function Determination Question, you have to maintain your perspective by considering for whom you are working. Since the manufacturing firm is funding this project by supplying its personnel to perform the study, you have to assume the role of the president or the stockholders when you answer this question.

Therefore, "make profit" becomes the basic function for this particular project at this point in time. We will discuss later how and why this may still change.

DEVELOP HOW TO MAKE PROFIT

Step 7. Develop Primary Path Functions

This step requires all the functions identified thus far to be listed, as shown in Figure 11.4. To accomplish this, we copy and paste the contents of Figures 11.2 and 11.3 and eliminate any duplicates.

Next, develop the primary path functions by asking the How Logic Question of the basic function and then asking the Why-How Logic Questions as more functions are added, as shown in Figure 11.5. We begin by posting the basic function on a separate line at the bottom of our working file, and then we ask:

> Q How is *make profit* actually accomplished?
> A Increase sales

Then we ask "why" and "how" of "increase sales" to see if it is properly placed in the FAST Tree. We repeat the process as new functions are posted. We continue the process as demonstrated in Figure 11.5 while at the same time

7	Heat Air
7	Cool Air
7	Dehumidify Air
7	Limit Deterioration
7	Minimize Repairs
	Simplify Mounting
7	Make Controls Accessible
7	Increase Efficiency
7	Make Product Attractive
7	Accommodate Normalcy
7	Supply Comfort
7	Satisfy Customer
7	Increase Sales
7	*Make Profit

Figure 11.4. List of Functions

```
Make Profit
  Increase Sales
    │ Attract Customers
    │     Improve Product
    │         Increase Efficiency
    │             Reduce Energy
    │     Make Product Attractive
    │         Produce Quality Product
    Satisfy Customer
    │     Manufacture Heat Pump Unit
    │     Supply Comfort
    │         Sell Heat Pump Unit
    │         Control Environment
    │             Heat Air
    │                 Generate Heat
    │                     Burn Gas
    │             Cool Air
    │                 Extract Heat
    │             Dehumidify Air
    │                 Remove Moisture
    │     Increase Product Life
    │         Limit Deterioration
    │             Prevent Rust
    │                 Protect Metals
    │                 Remove Moisture
    │             Withstand Elements
    │         Minimize Repairs
    │             Build Quality Product
    │     Make Controls Accessible
    │         Install Simple Covers
    │ Accommodate Normalcy
              Utilize Human Engineering Principles
```

Figure 11.5.   Heat Pump FAST Tree 1

trying to use any of the functions listed in Figure 11.4. If any of the functions in that list are merged into the FAST Tree, a "7" is placed near the left margin to indicate that they were used during Step 7. Note that 10 additional functions also have a "7" recorded in Figure 11.4, and many new functions have been added to Figure 11.5 as the How Logic Question was repeatedly asked.

FACILITATE INSTALLATION

Step 8. Evaluate Remaining Formulated Functions

Since one function, "simplify mounting," remains in Figure 11.4 which has not been merged into the logic diagram, it is copied and pasted at this point into the working file and the Why-How Logic Questions are asked of it to form the function cluster shown below. "Simplify mounting" is highlighted in bold as a reminder.

9	Facilitate Installation
10	**Simplify Mounting**
10	Simplify Connections
10	Utilize Modules
10	Analyze Construction

An investigation discloses that this function cluster does not fit into the primary path; therefore, we move on to Step 9.

SELL HEAT PUMP UNIT

Step 9. Using When/If Logic to Add Supporting Functions

As this step is performed, we investigate the primary path functions we have already identified and at the same time see if any clusters that remain in Step 8 can be merged into the logic diagram by asking the When/If Logic Question:

When/if *this function* is performed, what other functions must be performed?

A number of functions are inserted into this question until we come to the function "sell heat pump unit" without finding a place to locate the cluster developed during Step 8. When we insert this function, we get the following results:

Q1 When/if *sell heat pump unit* is performed, what other functions must be performed?
A1 Facilitate installation

This supporting function is added to the logic diagram, as shown in Figure 11.6, by placing a caret (^) symbol in front of it to indicate that it supports the

```
Make Profit
   Increase Sales
      Attract Customers
         Improve Product
            Increase Efficiency
               Reduce Energy
         Make Product Attractive
            Produce Quality Product
   Satisfy Customer
      Manufacture Heat Pump Unit
      Supply Comfort
         Sell Heat Pump Unit
         ^Facilitate Installation
            Simplify Mounting
            Simplify Connections
            Utilize Modules
               Analyze Construction
         Control Environment
            Heat Air
               Ignite Gas
            Cool Air
               Cool Refrigerant
            Dehumidify Air
               Remove Moisture
      Increase Product Life
         Limit Deterioration
            Prevent Rust
               Protect Metals
               Remove Moisture
            Withstand Elements
         Minimize Repairs
            Build Quality Product
      Make Controls Accessible
         Install Simple Covers
   Accommodate Normalcy
      Utilize Human Engineering Principles
```

Figure 11.6.   Heat Pump FAST Tree 2

function immediately above it. Also note that a "9" was placed near the left margin of the function "facilitate installation" of the function cluster listed during Step 8.

DEVELOP HOW TO FACILITATE INSTALLATION

Step 10. Develop Secondary Path Functions

The secondary path functions for this supporting function are added during this step, as shown in Figure 11.6. Four functions have already been identified in the function cluster above; however, when they are merged, the logic suggests that they be rearranged as shown. I suppose many more supporting functions could also be added. Since our main emphasis is always on creative ways of performing the primary path functions, normally not too much time is spent expanding the supporting branches of a logic diagram.

It is apparent that the logic still does not hold in both directions for the complete logic diagram shown in Figure 11.6. Therefore, the functions are rearranged until they do agree, as shown in FAST Tree 3 in Figure 11.7.

A check of the Why-How Logic of this third FAST Tree shows that it seems to hold in both directions quite well. Since the intent of this project is to improve the existing three-ton heat pump, the remaining portion of our analysis is devoted to the branch of the FAST Tree that pertains to the task we were assigned, namely "improve existing heat pump unit." Therefore, the remainder of this sample customer-oriented project will deal with FAST Tree 4 shown in Figure 11.8.

Since products selected for this type of analysis have already been built and are normally being sold to customers, clients, or consumers, a team of four to six people is usually selected to perform the study. Such an effort is usually performed with all participants working together in the same room. A FAST Diagram is the standard method to proceed right from the start. However, I recommend that a FAST Tree be constructed using a computer at the same time the FAST Diagram is constructed because a word-processing program can keep track of all your thinking and analysis for you and at the same time make sure you have considered all functions you have formulated.

It is easy to convert a FAST Tree into a FAST Diagram with the Visio® templates available on the publisher's web site (www.jrosspub.com). Appendix A describes in detail how to construct FAST Diagrams from FAST Trees. FAST Tree 4 shown in Figure 11.8 is converted into the FAST Diagram shown in Figure 11.9.

As you can see, it took some time to filter out how we wanted to proceed with this project. Perhaps a team might want to bypass the intense thinking that is required to reach this same basic function, namely "improve existing heat pump unit." A team may want to start with "improve existing unit" and replace "existing unit" with the name of the product. It is not a bad approach, but I would suggest that Steps 1 through 6 all be performed before making that

```
Make Profit
  Increase Sales
    Attract Customers
      Invent Heat Pump Unit
        Control Environment
          Heat Air
            Ignite Gas
          Cool Air
            Cool Refrigerant
          Dehumidify Air
            Remove Moisture
      Improve Existing Heat Pump Unit
        Increase Efficiency
          Reduce Energy
        Make Product Attractive
          Produce Quality Product
        Increase Product Life
          Limit Deterioration
            Prevent Rust
              Protect Metals
              Remove Moisture
            Withstand Elements
          Minimize Repairs
            Build Quality Product
        Make Controls Accessible
          Install Simple Covers
        Facilitate Installation
          Simplify Mounting
          Simplify Connections
          Utilize Modules
            Analyze Construction
        Accommodate Normalcy
          Utilize Human Engineering Principles
    Satisfy Customers
      Supply Comfort
        Sell Heat Pump Units
```

Figure 11.7. Heat Pump FAST Tree 3

decision; otherwise, some important facts that could greatly improve your project may not come to light. Once the final basic function has been selected, the team should make sure the Why-How Logic holds throughout the entire FAST Diagram. This is best verified by asking the following Verification Question:

Improve Existing Heat Pump Unit
  Increase Efficiency
    Reduce Energy
  Make Product Attractive
    Produce Quality Product
  Increase Product Life
    Limit Deterioration
      Prevent Rust
        Protect Metals
        Remove Moisture
      Withstand Elements
    Minimize Repairs
      Build Quality Product
  Make Controls Accessible
    Install Simple Covers
  Facilitate Installation
    Simplify Mounting
    Simplify Connections
    Utilize Modules
      Analyze Construction
  Accommodate Normalcy
    Utilize Human Engineering Principles

Figure 11.8. Heat Pump FAST Tree 4

Does *this how function* help *its why function*?

A few examples of Verification Questions that should be asked for the FAST Diagram shown in Figure 11.9 and their corresponding answers are as follows:

Q   Does *protect metals* help *prevent rust*?
A   Yes

Q   Does *remove moisture* help *prevent rust*?
A   Yes

Q   Does *prevent rust* help *limit deterioration*?
A   Yes

How ➝   ⬅ Why

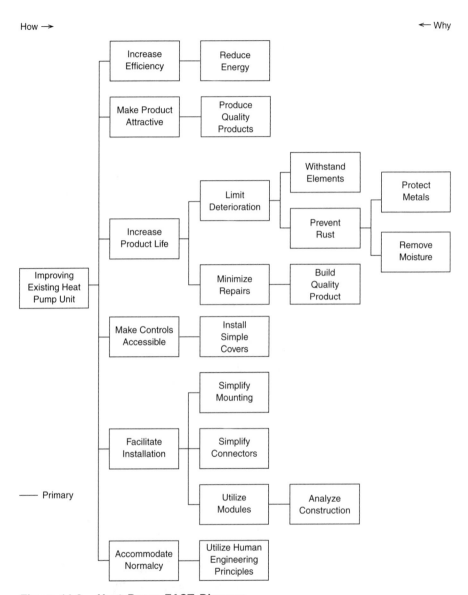

Figure 11.9.   Heat Pump FAST Diagram

Q   Does *withstand elements* help *limit deterioration*?
A   Yes

Q   Does *limit deterioration* help *increase product life*?
A   Yes

This process is continued until every how function has been checked, which only takes a few minutes. If any question yields a "no" answer, then that how function should be changed until it describes how it helps the why function. Other methods of helping the why function should also be creatively explored and added to the diagram. Once this has been completed, those functions that have the greatest potential for improving the end product should then be explored. The functions that have the greatest potential for improving the end product are what the team is actually seeking during this entire function analysis exercise.

IMPROVE EXISTING HEAT PUMP UNIT

Step 11. Brainstorming Higher Level Functions

The first question we ask when we brainstorm is:

How else can *this function* be accomplished or performed?

We will explore just the following three higher level functions to demonstrate how to use this How Else Question to expand our thinking and stimulate our creativity on this type of project:

> Improve Existing Heat Pump Unit
> Increase Efficiency
> Increase Product Life

As the following How Else Questions are asked, the same ground rules outlined in Chapter 8 are required for performing brainstorming. Note that instead of writing our answers as sentence statements, they are recorded directly as functions. The results gleaned from this creative exercise are incorporated into FAST Tree 5 shown in Figure 11.10.

Q1   How else can *improve existing heat pump unit* be accomplished or performed?

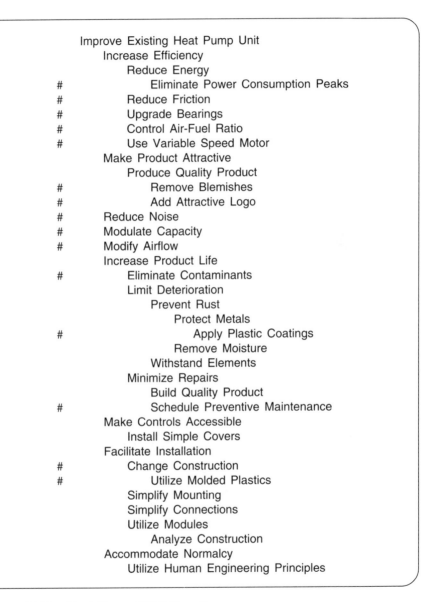

```
                    Improve Existing Heat Pump Unit
                        Increase Efficiency
                            Reduce Energy
        #                       Eliminate Power Consumption Peaks
        #                   Reduce Friction
        #                   Upgrade Bearings
        #                   Control Air-Fuel Ratio
        #                   Use Variable Speed Motor
                        Make Product Attractive
                            Produce Quality Product
        #                       Remove Blemishes
        #                       Add Attractive Logo
        #               Reduce Noise
        #               Modulate Capacity
        #               Modify Airflow
                        Increase Product Life
        #                   Eliminate Contaminants
                            Limit Deterioration
                                Prevent Rust
                                    Protect Metals
        #                               Apply Plastic Coatings
                                    Remove Moisture
                                Withstand Elements
                            Minimize Repairs
                                Build Quality Product
        #                           Schedule Preventive Maintenance
                        Make Controls Accessible
                            Install Simple Covers
                        Facilitate Installation
        #                   Change Construction
        #                       Utilize Molded Plastics
                            Simplify Mounting
                            Simplify Connections
                            Utilize Modules
                                Analyze Construction
                        Accommodate Normalcy
                            Utilize Human Engineering Principles
```

Figure 11.10.   Heat Pump FAST Tree 5

Role   Engineer

A1-1   Reduce noise, modulate capacity, modify airflow, and change construction

Role Artist
A1-2 Remove blemishes, include attractive logo

Q2 How else can *increase efficiency* be accomplished or performed?

Role Mechanical engineer
A2-1 Reduce friction, upgrade bearings, control air-fuel ratio

Role Electrical engineer
A2-2 Use variable speed motor, eliminate power consumption peaks

Q3 How else can *increase product life* be accomplished or performed?

Role Engineer
A3-1 Utilize molded plastics and apply plastic coatings

Role Metal part
A3-2 Eliminate contaminants

Role Doctor
A3-3 Schedule preventive maintenance

LOGIC DIAGRAMS: A SPRINGBOARD TO CREATIVITY

We have incorporated all the new functions we generated during our brain-storming session and have indicated them by a pound (#) symbol near the left margin. Obviously, many more could be added if we applied our brainstorming technique to the other functions in the FAST Diagram. It can be said that any logic diagram may be used as a springboard to creativity.

Once a logic diagram has been constructed, it provides many additional opportunities to be creative by asking the How Logic Question of each function within the diagram. The primary benefit of the FAST technique is that it expands your understanding of your project and helps to collect any information that may aid you in arriving at a creative and worthwhile solution or proposal. Make sure you take the time to expand the various function clusters during Step 8. It will soon become obvious where these clusters can be merged into your logic diagram. It may still require considerable deep thinking by asking the Why Logic Question of each cluster over and over again to make sure an important cluster has not been eliminated that should be included. Once you have a FAST Diagram like the one shown in Figure 11.9, you have many areas you can work on to expand your understanding of your project and to stimulate your creativity.

Note that in the example in this chapter, we went through a lot of functions which helped us decide where we should concentrate our final effort. You never know which function you identify may trigger your most creative work. I limited the number of roles that could have been played as we brainstormed the higher level functions of this project. I trust that those who analyze consumer-type products will find this example to be of some value.

12

PROJECT 5: MILITARY COMMUNICATION DEVICE*

This chapter demonstrates how to analyze a process, procedure, piece of equipment, or anything else that has one or more problems. The main objective of this type of project is to solve existing problems and at the same time improve the process, procedure, or equipment. The project selected for this chapter is a piece of equipment with several problems that need to be fixed. Also, this equipment has to be reliable because in some instances it can be a matter of life and death if it fails to function properly. This chapter also demonstrates how to indicate alternate proposals on a logic diagram. Method 5 is usually chosen for selecting this type of project.

---

* The basic information for this project, the team selection, initial analysis, and naming of functions was supplied by Jerry J. Kaufman, President, J.J. Kaufman Associates, Inc., Houston, Texas.

USING METHOD 5

Step 1. Selecting a Project

The project selected is a communication device used by the military to display, analyze, and transmit field conditions. The device is a portable electronics package easily carried and inserted into and removed from a variety of host vehicles. The main components are a structural case that contains the system's electronics, a removable hard drive, and backup battery. Also included in the device are a display, keyboard, and interface cables.

Method 5 requires four questions to be asked, which yielded the following answers:

Q1   What problem shall we discuss?

A1   Current communication system has acquisition and sustainment costs that exceed projected annual acquisition plus ownership allocations

Q2   Why do you think this is a problem?

A2   Field failures have exceeded normal operating costs
- Water intrusion in the power unit reduces operational life
- Rough handling causes bezel button failure
- Interfaces between components degrade reliability
- Vibration causes electrical connectors to disengage
- System lockups and rebooting occur too frequently

Q3   Why do you think a solution is needed?

A3   Continued high cost will affect replacement quantities

Q4   What is there about this problem area that disturbs you?

A4   Failure could lead to unacceptable field performance

A PRODUCT DESIGN TEAM AND A PRODUCT USER TEAM

Step 2. Selecting Participants

At this point, it appears that the problem resolution needs to concentrate on the product's design and the handling and operation of the equipment in the field. That being the case, the potential solution suggests that two teams be formed to analyze this project. One team, representing the contractor, is to focus on the design of the system. A second team, consisting of users of the product, who are army field personnel, is to address the operation and maintenance of the product. Both teams will work within the same room.

CONTROL ACQUISITION COSTS

Step 3. Initial Functions

In order to create a list of functions, we must convert the answers developed during Step 2 into functions. The answers are as follows:

A1   Current system has acquisition and sustainment costs that exceed projected annual acquisition plus ownership allocations

A2   Field failures have exceeded normal operating costs
- Water intrusion in the power unit reduces operational life
- Rough handling causes bezel button failure
- Interfaces between components degrade reliability
- Vibration causes electrical connectors to disengage
- System lockups and rebooting occur too frequently

A3   Continued high cost will affect replacement quantities

A4   Failure leads to unacceptable field performance

The initial functions gleaned from these answers are shown in Figure 12.1.

Control Acquisition Costs
Control Sustainment Costs
Eliminate Water Intrusion
Prevent Bezel Button Failures
Increase Interface Reliability
Redesign Connectors
Eliminate Program Memory Leaks
Eliminate Modifying Operating System Kernel Calls
Ensure Line Replaceable Unit (LRU) Availability
*Increase Field Performance

Figure 12.1.   Initial Functions

INCREASE FIELD PERFORMANCE

Step 4. Initial Basic Function

From the list in Figure 12.1, it was determined that "increase field performance" is probably the initial basic function. Therefore, this function is inserted into the Basic Function Determination Question:

Q   If *increase field performance* didn't have to be performed, would any of the other functions still have to be performed?

A   If we assume that the performance will increase enough so acquisitions will be reduced and sustainment costs will also be reduced, then the response to this question is no for all other initial functions.

Therefore, our *initial basic function* is indeed *"increase field performance."*

UTILIZE HIGHER LEVEL LOGIC QUESTIONS

Step 5. Develop Higher Level Functions

Looking for higher level functions requires asking and answering the following three questions as we role-play:

Q1   Why is it necessary to *increase field performance*?

Role   Officer in charge of communication device
A1   Vital required information fails to be transmitted and received

Q2   What higher level function caused *increase field performance* to come into being?

Role   Officer in charge of an entire battalion
A2   The higher level function is "communicate between units"

Q3   What is really trying to be accomplished when *increase field performance* is performed?

Role   Officer in charge of an entire battalion
A3   We are trying to support the mission assigned to us and other units in any combat situation

Increase Field Communication Performance
Transmit Vital Information
Receive Vital Information
Communicate Between Units
*Support Mission

Figure 12.2. Higher Level Functions

We glean from these answers the higher level functions listed in Figure 12.2, and we included in this list our initial basic function.

SUPPORT MISSION

Step 6. Identify the Basic Function

Note that we added the modifier "communication" to the first function in our list in order to intelligently ask the Basic Function Determination Question. If we assume "support mission" is the basic function of these five functions, we next ask the question:

> Q   If *support mission* didn't have to be performed, would any of the other functions still have to be performed?

> A   No

Therefore, "*support mission*" becomes the *basic function.*

DEVELOP HOW TO SUPPORT MISSION

Step 7. Develop Primary Path Functions

The first task is to list all the functions generated thus far, as shown in Figure 12.3. Next, we develop the primary path functions as much as we can by asking the Why-How Logic Questions. As we do this, we try to fit the functions listed in Figure 12.3 into the FAST Tree.

Since we were unable to develop the FAST Tree more than what is shown in Figure 12.4 during our first cut at it, we move on to Step 8.

Increase Field Communication Performance
Transmit Vital Information
Receive Vital Information
Communicate Between Units
Control Acquisition Costs
Control Sustainment Costs
Eliminate Water Intrusion
Prevent Bezel Button Failures
Increase Interface Reliability
Redesign Connectors
Eliminate Program Memory Leaks
Eliminate Modifying Operating System Kernel Calls
Ensure LRU Availability
7   *Support Mission

Figure 12.3. List of Functions

Support Mission
Determine Response

Figure 12.4. Support Mission FAST Tree 1

DEPLOY RELIABLE COMMUNICATION EQUIPMENT

Step 8. Evaluate Remaining Formulated Functions

We post all the remaining functions by copying and pasting them two or three tab spaces from the left margin at the end of our working file. Then we separate them so we can apply the Why-How Logic Questions to each function, as shown below. We also highlight in bold only the lines that have functions on them so we can remember which functions were in our original list before the clusters are expanded. Since the list of primary path functions developed thus far appears to be incomplete, it is obvious that we know very little about this project at this stage of development of the logic diagram. Frequently it is necessary as these clusters are being developed to repeatedly ask the Why Logic Question until we can fit the clusters into the primary path. This is often the case when a complicated project is analyzed.

7	Support Mission
8	Deploy Reliable Communication Equipment
8	**Increase Field Communication Performance**
8	Eliminate Failures
7	Determine Response
8	Share Tactical Intelligence
8	**Transmit Vital Information**
8	Operate Communication System
7	Determine Response
8	Process Information
8	Obtain Tactical Intelligence
8	**Receive Vital Information**
8	Operate Communication System
8	Process Information
8	Obtain Tactical Intelligence
8	**Communicate Between Units**
8	Transmit Vital Information
8	Operate Communication System
8	Maintain Inventory
8	**Control Acquisition Costs**
8	Repair Faulty LRUs
8	Test LRUs
8	Replace Faulty Components
8	Increase Field Communication Performance
9	**^Control Sustainment Costs**
10	Service Equipment
10	Stock Spare LRUs
8	Preserve Environmental Integrity
8	Exclude Contaminants
8	**Eliminate Water Intrusion**
8	Seal LRUs
8	Deploy Reliable Communication Equipment
8	**Prevent Bezel Button Failures**
8	Redesign Bezel Buttons

7   Support Mission
8   Deploy Reliable Communication Equipment
8   Increase Field Communication Performance
8   **Increase Interface Reliability**
8   Improve Connectors

8   Improve Connectors
8   **Redesign Connectors**
8   Prevent Vibration Disengagement

Note that we have checked each function cluster to see if any supporting functions are included within a cluster by asking the following question:

Does *this how function* help *its why function*?

We only found one supporting function: "control sustainment costs." This suggests that the other functions will probably merge into the primary path. Note that a "7" has been added near the left margin for the two functions included during Step 7. An "8" is added for all functions merged during this step in the procedure.

As the function clusters were merged into FAST Tree 2 shown in Figure 12.5, it became obvious that the logic did not always agree from cluster to cluster. This is what usually happens. When it does, you have to cut, relocate, and paste the functions until the logic holds in both directions within your logic diagram. Also, the names of functions often have to be changed to agree with the consensus of the group. For example, the function "share tactical intelligence" appeared to be better than "obtain tactical intelligence."

CONTROL SUSTAINMENT COSTS

Step 9. Using When/If Logic to Add Supporting Functions

Q   When/if *this function* is performed, what other functions must be performed?

A   Supporting functions are preceded by a caret (^) symbol as they are merged into the FAST Tree, as shown in Figure 12.6.

Usually, if we are after a creative solution, we would concentrate our efforts at this stage of our analysis on the primary path functions shown in Figure 12.6.

Support Mission
  Deploy Reliable Communication Equipment
    Increase Field Communication Performance
      Prevent Bezel Button Failures
        Redesign Bezel Buttons
      Preserve Environmental Integrity
        Exclude Contaminants
          Eliminate Water Intrusion
            Seal LRUs
      Increase Interface Reliability
        Improve Connectors
          Redesign Connectors
            Prevent Vibration Disengagement
      Eliminate Failures
        Observe Anomaly
          Train Communication Personnel
            Train Operators
            Train Maintenance Personnel
        Remove Faulty LRUs
          Isolate Faults
            Test System
              Activate Built-in Test
  Maintain Inventory
    Control Acquisition Costs
      Repair Faulty LRUs
        Test LRUs
        Replace Faulty Components
      Acquisition New LRUs
      Scrap LRUs
  Determine Response
    Process Information
      Share Tactical Intelligence
        Communicate Between Units
          Operate Communication System
            Select Bezel Button
            Transmit Vital Information
            Input Information
              Sensitize Screen
              Activate Mouse
              Initiate Keyboard
            Receive Vital Information

Figure 12.5. Support Mission FAST Tree 2

Support Mission
    Deploy Reliable Communication Equipment
        Increase Field Communication Performance
           Prevent Bezel Button Failures
              Redesign Bezel Buttons
           Preserve Environmental Integrity
              Exclude Contaminants
                  Eliminate Water Intrusion
                      Seal LRUs
        Increase Interface Reliability
           Improve Connectors
              Redesign Connectors
                  Prevent Vibration Disengagement
        Eliminate Failures
           Observe Anomaly
              Train Communication Personnel
              Train Operators
              Train Maintenance Personnel
           Remove Faulty LRUs
           Isolate Faults
              Test System
                  Activate Built-in Test
      ^Control Sustainment Costs
    Maintain Inventory
        Control Acquisition Costs
           Repair Faulty LRUs
              Test LRUs
              Replace Faulty Components
           ^Confirm LRU Anomaly
        Acquisition New LRUs
        Scrap LRUs
    Determine Response
        Process Information
           Share Tactical Intelligence
              Communicate Between Units
                  Operate Communication System
                  Select Bezel Button
                  Transmit Vital Information
                  Input Information
                      Sensitize Screen
                      Activate Mouse
                      Initiate Keyboard
                ^Display Data
              Receive Vital Information
        ^Eliminate Computer Lockups

Figure 12.6. Support Mission FAST Tree 3

Since the construction of a FAST Diagram for all the functions listed in FAST Tree 3 is the best way for a team of participants to work together, FAST Tree 3 for "support mission" is converted into the FAST Diagram shown in Figure 12.7.

This diagram allows the relationships between functions to be easily visualized and understood as several participants look at this logic diagram. Once again, we can pick any function posted on this drawing and ask "Why must this function be performed?" The answer appears in the function block connected at its left. Likewise, we can ask "How is this function actually performed or proposed to be performed?" This answer appears in all the function blocks connected at its right.

DEVELOP HOW TO ELIMINATE COMPUTER LOCKUPS

Step 10. Develop Secondary Path Functions

Once the secondary path functions have been added to FAST Tree 4 as shown in Figure 12.8, the logic should be checked to see if it holds in both directions. When this is accomplished, a complete stranger looking at FAST Tree 4 can become familiarized with the project within minutes. This was forcefully brought to my attention several years ago when I was working for the Sperry Rand Corporation as we were developing some mail sorting equipment for the United States Post Office.

We decided to analyze an existing pneumatic switching device using my technique. As we analyzed this switching device and identified the functions performed by it, we proceeded to ask the Why-How Logic Questions, which are the meat and heart of this technique; they determine why and how each function is performed. Within a few hours, we had an entirely new concept for a pneumatic switching device. As we proceeded to reduce this concept into hardware, we included the machinist and the technician in our discussions. This was perhaps the first time these two individuals knew why engineers wanted to perform a given function and how we planned on reducing our concept into actual hardware. The why function in the logic diagram informed these two individuals why we wanted that function performed in the first place, and the how function told them how we physically proposed to design and build this newly mentally conceived switching device.

When our first design did not work, we asked our machinist to machine another block of metal of approximately the same dimensions but with a few minor variations. The machinist told me that he had planned on leaving the company because he did not think the engineers could make up their minds about what they really wanted. As soon as he knew the whys and hows of this

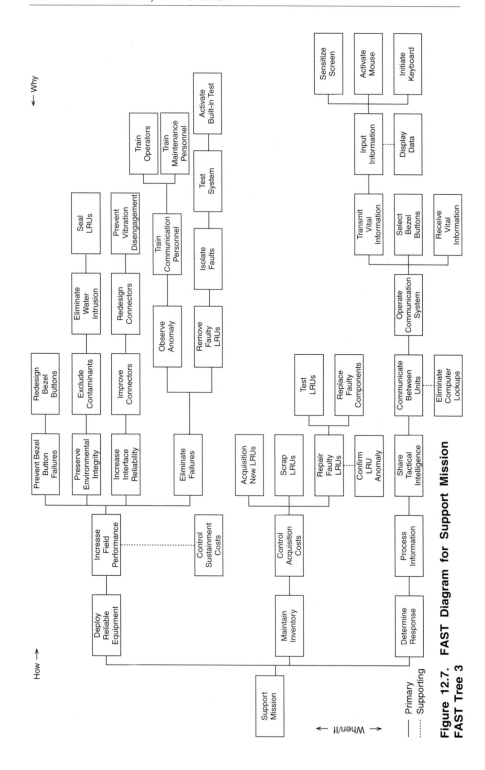

Figure 12.7.   FAST Diagram for Support Mission
FAST Tree 3

```
Support Mission
    Deploy Reliable Communication Equipment
        Increase Field Communication Performance
            Prevent Bezel Button Failures
                Redesign Bezel Buttons
            Preserve Environmental Integrity
                Exclude Contaminants
                    Eliminate Water Intrusion
                        Seal LRUs
            Increase Interface Reliability
                Improve Connectors
                    Redesign Connectors
                        Prevent Vibration Disengagement
            Eliminate Failures
                Observe Anomaly
                    Train Communication Personnel
                        Train Operators
                        Train Maintenance Personnel
                Remove Faulty LRUs
                    Isolate Faults
                        Test System
                            Activate Built-in Test
        ^Control Sustainment Costs
            Service Equipment
            Stock Spare LRUs
    Maintain Inventory
        Control Acquisition Costs
            Repair Faulty LRUs
                Test LRUs
                Replace Faulty Components
            ^Confirm LRU Anomaly
                Verify Failure Exists
            Acquisition New LRUs
            Scrap LRUs
    Determine Response
        Process Information
            Share Tactical Intelligence
                Communicate Between Units
                    Operate Communication System
                        Select Bezel Button
                        Transmit Vital Information
                            Input Information
                                Sensitize Screen
```

Figure 12.8.   Support Mission FAST Tree 4

Activate Mouse
Initiate Keyboard
^Display Data
Receive Vital Information
^Eliminate Computer Lockups
Sequence Shutdown
Instruct Operator
Assure Power
Store Emergency Energy
Document Lockups
Identify Program Running
Identify Steps Performed
Perform Quality Assurance on Programs
Use Structure Programming
Eliminate Program Memory Leaks
Account for All Memory Used
Deallocate Used Allocated Memory
Close Programs
Close Routines
Eliminate Viruses
Eliminate Modifying Operating System Kernel Calls
Check Program Calls to Operating System

Figure 12.8 continued.

project, his work took on new meaning, and he decided to stay with the company as a result of his involvement in the design. As the project progressed, he offered several suggestions that affected the design. After our second or third modification, this pneumatic device started to function as we had planned.

After a few moments of operation, however, it was discovered that the switching was only temporary. The main air jet stream began switching on and off with an apparent uniform frequency. This particular study resulted in the discovery of a new phenomenon. Scientists, engineers, and physicists who had worked in this field of research were surprised to learn that this phenomenon existed. When it was realized that it was new, it was named the Fox-Thorne Phenomenon, after the two engineers who performed the experiments in the laboratory. Perhaps it would not have been discovered without the suggestions of this machinist who became an active participant in the project. Informed personnel are always more productive, and everyone is willing to contribute if allowed to help.

Let's return to our communication project. Now that our FAST Tree for this communication system has been created, we need to decide what portion or branch of the diagram each team should be responsible for. Once that has been decided, a search is made to select functions to be explored for further creativity. In this particular case, each of the teams selected five different branches of the tree to explore and make proposals for solving the problems and improving the system. Since we are not privy to their creative effort, I have selected various functions to explore in order to demonstrate how to proceed after the initial logic diagram has been created. Note that many creative ideas were already disclosed as the diagram was being expanded.

REDESIGN BEZEL BUTTONS AND OBSERVE ANOMALY

Step 11. Brainstorming Higher Level Functions

Brainstorming when using a function as the focal point is best accomplished by role-playing when the How Else Logic Question is asked:

How else can *this function* be accomplished or performed?

The function inserted into the question can be any function listed in the FAST Tree, but usually one or more of the higher level functions on the primary path are selected. For this example, we will limit our investigation to the following three functions:

> Redesign Bezel Button
> Observe Anomaly
> Sequence Shutdown

The questions we need to answer as we brainstorm these functions are listed below as Q1, Q2, and Q3. Since more than one answer often comes to mind, the answers are listed as A1 through A3, followed by a hyphen and a separate number for each answer. The proposals from this investigation are listed as proposals P1-1 through P3-2. Each hyphenated number represents a separate proposal. Since these are proposals, they are preceded by a question mark (?) when they are merged into the FAST Tree, which indicates that only one of each proposal number should be implemented into the final redesigned product.

Q1   How else can *redesign bezel button* be accomplished or performed?

This function suggests that the present design be changed; however, in attempting to answer this question, one should think of the functions performed by the bezel buttons. People familiar with computers might think of the following:

A1-1   Place icons on a computer toolbar which when clicked by a mouse will cause the same things to occur as when a bezel button is selected.

A1-2   Use a touch screen that has an area titled "Bezel Buttons." Have all possible buttons come up so any one of them can be touched in order to carry out the desired selection.

A1-3   Use toggle switches instead of buttons, with up for "on" and down for "off." Seal the row of toggle switches with a rubber boot so they are not exposed to the outside environment.

Now that we have answered the questions, we need to transform our answers into proposals. We do this by creating function clusters for each proposal:

P1-1   Utilize Screen Icons
        Generate Screen Icons
            Perform Programming
        Activate Screen Icons
            Click Icons On/Off

P1-2   Implement Touch Screen
        Generate Main Button Area
        Sensitize Main Button Area
        Generate Button Function Areas
        Sensitize Button Function Areas
        Touch Desired Function Area

P1-3   Install Toggle Switches
        Activate Toggle Switches
        Seal Toggle Switches
            Mold Switch Boot

Q2   How else can *observe anomaly* be accomplished or performed?

A2-1   Store standard responses in memory and compare each response to that standard. If a response fails to agree with the standard, have the computer record and store the anomaly. Use this information to correct the problem either by reprogramming or by replacing faulty LRUs.

A2-2 Have the computer's software sequence keys to be pushed during each routine, so when an anomaly occurs that sequence can be stored. An anomaly can then be duplicated when different LRUs are tested so the cause of the anomaly can be corrected.

P2-1 Observe Computer Anomaly
    Store Standard Response
    Compare Routine with Standard
    Store Anomaly Deviations
    Correct Problems
    Correct Faulty Programs
    Replace Faulty LRUs

P2-2 Correct Anomaly Failures
    Document Computer Anomaly
        Document Key Sequence
        Store Anomaly Sequences
        Test LRUs
        Use Stored Anomaly Sequences

Q3 How else can *sequence shutdown* be accomplished or performed?

A3-1 Have computer shutdown buttons, icons, toggles, or touch screen areas that when activated allow the computer to automatically designate and execute the steps that must be taken to shut down. Programs thus designed will shut down the computer correctly every time.

A3-2 Use interlocks to regulate shutdown sequence. Once the first step in the sequence has been performed, the interlocks prevent the other steps from being performed in the wrong sequence.

P3-1 Shutdown System
    Activate First Step
    Computerize Other Steps

        P3-2   Shutdown System
                Utilize Interlocks
                Send Signals to AND Gates

PROPOSALS MUST BE DEVELOPED

Obviously, these are just proposals which must be developed and evaluated. I have no idea what was proposed by the two teams and am merely showing what might have taken place as these teams performed their brainstorming sessions. A method for incorporating these proposals into a FAST Tree is shown in Figure 12.9. Note that as the proposals were being conceived, function clusters were generated that expressed how each proposal might be implemented. Each of the proposals opens new doors for expanding a person's creativity.

    Note that each function we inserted into the How Else Questions is also identified in the FAST Tree by placing a question mark (?) and a proposal

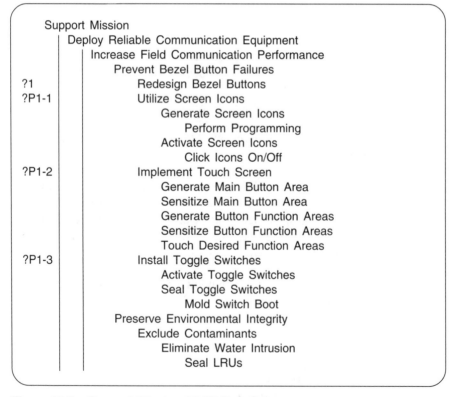

Figure 12.9. Support Mission FAST Tree 5

```
                                Increase Interface Reliability
                                   Improve Connectors
                                      Redesign Connectors
                                         Prevent Vibration Disengagement
                                Eliminate Failures
                                   Observe Anomaly
?2                                    Train Communication Personnel
                                         Train Operators
                                         Train Maintenance Personnel
?P2-1                                 Observe Computer Anomaly
                                         Store Standard Response
                                         Compare Routine with Standard
                                         Store Anomaly Deviations
                                         Correct Problems
                                            Correct Faulty Programs
                                            Replace Faulty LRUs
?P2-2                                 Correct Anomaly Failures
                                         Document Computer Anomaly
                                            Document Key Sequence
                                            Store Anomaly Sequences
                                            Test LRUs
                                               Use Stored Anomaly Sequences
                                   Remove Faulty LRUs
                                      Isolate Faults
                                         Test System
                                            Activate Built-in Test
                          ^Control Sustainment Costs
                             Service Equipment
                             Stock Spare LRUs
                  Maintain Inventory
                     Control Acquisition Costs
                        Repair Faulty LRUs
                           Test LRUs
                           Replace Faulty Components
                        ^Confirm LRU Anomaly
                           Verify Failure Exists
                        Acquisition New LRUs
                        Scrap LRUs
                  Determine Response
                     Process Information
                        Share Tactical Intelligence
                           Communicate Between Units
                              Operate Communication System
```

Figure 12.9 continued.

```
                              Select Bezel Button
                           Transmit Vital Information
                              Input Information
                                 Sensitize Screen
                                 Activate Mouse
                                 Initiate Keyboard
                              ^Display Data
                           Receive Vital Information
                      ^Eliminate Computer Lockups
                           Document Lockups
                              Identify Program Running
                              Identify Steps Performed
?3                         Sequence Shutdown
                              Instruct Operator
                              Assure Power
                                 Store Emergency Energy
?P3-1                      Shutdown System
                              Activate First Step
                              Computerize Other Steps
?P3-2                      Utilize Interlocks
                                 Send Signals to AND Gates
                           Perform Quality Assurance on Programs
                           Use Structure Programming
                           Eliminate Program Memory Leaks
                              Account for All Memory Used
                              Deallocate Used Allocated Memory
                              Close Programs
                              Close Routines
                           Eliminate Viruses
                           Eliminate Modifying Operating System Kernel
                           Calls
                              Check Program Calls to Operating System
```

Figure 12.9 continued.

number next to it. This method is used to identify the functions we brainstormed. This allows us to show alternate methods of performing the same functions immediately below one another, as illustrated in Figure 12.9 for proposals P1-1 through P3-2.

READING A FAST TREE

You may have noticed that the words "how" and "why" and the accompanying arrows are not included in the FAST Trees in this chapter. By this time, you should be well indoctrinated in the procedure and know that why functions are always above and one tab space to the left of any function considered and how functions are always below and one tab space to the right of any why function. Since more than one how function may describe how a given function is performed or proposed to be performed, how functions are always positioned at the same tab space from the left margin. The sequence in which these functions are to be listed in a logic diagram is optional; however, they are usually listed in the sequence in which they should be considered.

When a FAST Tree is being shown to others who are not familiar with this technique, it is always a good idea to show the "how" and "why" notations along with the arrows. When a logic diagram is being presented to a large audience, it is much more impressive and informative if displayed as a FAST Diagram. PowerPoint® presentations of FAST Trees and FAST Diagrams are perhaps the most effective way to display the information to a large group of people. The names of the functions should be large enough for everyone to read. Always explain to any group viewing a logic diagram the enormous amount of information contained within the diagram by picking out two or three important functions and showing the audience why each function is necessary by referring to the function at its immediate left. Also show how each function is accomplished or proposed to be accomplished by referring to one or more functions at its immediate right. These functions are always connected by a line from one function to the next in a FAST Diagram. Refer to Appendix A when constructing a FAST Diagram.

13

GENERALIZING AND UNDISCLOSING METHODS

Steps 12 and 13 of the FAST procedure are presented in this chapter. Step 12 demonstrates how generalizing functions can open new doors to creativity, thereby removing any creative limitations that may exist because of the names given to certain functions. Then Step 13 is presented; it outlines how to formulate a new function that does not disclose the method described by a recorded function in your completed logic diagram. This likewise can stimulate additional opportunities to be creative.

GENERALIZING FUNCTIONS

If you and your companion are invited to a neighborhood party, the host of the party might suggest that you bring some apples. The two of you then have to decide what kind of apples you will bring. What function do you have to perform in order to comply with the host's request? See if you can describe the function using an active verb and a measurable noun. One obvious function is "bring apples." Is the word "apples" a measurable noun? Let's analyze it for a moment. You can bring a bushel, a peck, or perhaps just a small paper sack full of apples. Apples can be small, a couple of inches in diameter, or larger. Thus it is obvious that this noun is measurable.

But apples have other measurable qualities, such as variety, taste, and even color. The variety of apples you choose, how they taste, and what color to select require decisions to be made. The host could have suggested that you bring six Golden Delicious Apples. This suggestion would have limited the number of decisions you have to make, which in turn would have limited your opportunity to be creative. Even the number of apples to bring has been designated. Every time you have to make a decision, you have an opportunity to be creative.

Suppose your host had suggested that you bring some fruit. Now your function becomes "bring fruit." Then you have to decide what kind of fruit. Should it be two or three different kinds, different colors, different shapes, or what? The host could have suggested a salad, a dessert, or just bring food. Then your functions would become "bring salad," "bring dessert," or "bring food." These first two limit you a little bit but provide a lot more opportunity for creativity than "bring apples." The function "bring food" opens the door almost wide open. This demonstrates that there exists a greater opportunity to be creative when a function is generalized. Therefore, try to generalize your functions when you are seeking additional creative opportunities.

When you start to write functions, always feel free to use one or more modifiers if you need to. You might want to add modifiers initially so you don't forget a concept or thought. If the host in the above example had requested that you bring Golden Delicious Apples, you could have mentally thought of the function "bring Golden Delicious Apples." Thinking in terms of functions makes tasks easier to remember. Functions that contain modifiers are sometimes called *apparent functions* because they contain more than a verb and a noun; some people believe modifiers limit creativity. You may have noticed that I rarely formulate a function that does not have a modifier. I have not found that this practice stifles my creativity. This may be because I always role-play when I try to be creative.

The first function listed in Figure 13.1 is very specific and does not require a great amount of thinking or creativity to perform. As you move down the list, the creative opportunities continue to be expanded. This method of increasing your creative opportunities is called *generalizing functions*.

The exercise of trying to generalize a function can open many new creative doors for you. Many times, all you have to do to generalize a function is remove the modifier. Modifiers, on the other hand, are especially helpful when you are starting to collect information about a project, because they have a tendency to make things more specific, which helps people remember concepts better. Take a few minutes and generalize the functions listed in Figure 13.2. Doing so will require some deep thinking on your part.

You probably discovered that this exercise is not as easy as the first example with Golden Delicious Apples. The list in Figure 13.3 shows how I generalized

Bring Golden Delicious Apples
Bring Delicious Apples
Bring Apples
Bring Fruit
Bring Food
Bring Substance
Bring Anything

Figure 13.1. Generalizing Functions

these functions. Sometimes just a small amount of generalization will aid you in your creativity.

GOLDEN DELICIOUS APPLES VERSUS FRUIT

Step 12. Generalizing Functions

As you read one of the initial functions in Figure 13.3, your mind takes you one place, and when you read its generalized function, it takes you another place. That's what creativity is all about. After you have generated all the ideas you can by performing Steps 1 though 11, perform this step by scanning the functions in your FAST Tree or FAST Diagram and see if you can generalize any of them. Creative ideas come to people because they devote time and energy to searching for them. I believe that if you try to make this world a better place in which to live, you will always be creative. I know a man who is always thinking in areas where others do not venture. I trust that the concepts within this book will help you do the same.

Figure 13.4 shows five functions gleaned from the logic diagrams used to demonstrate how to apply the first 11 steps of the FAST procedure presented

Attract Attention	Collect Dirt	Control Deflection
Convert Energy	Create Image	Educate Students
Suppress Noise	Cut Grass	Prevent Contamination
Remove Heat	Tighten Bolt	Transport People
Heat Filament	Eat Ham	Eat Toast
Reduce Friction	Filter Noise	Control Anger

Figure 13.2. Ungeneralized Functions

Initial Functions	Generalized Functions
Attract Attention	Stimulate Senses
Convert Energy	Change Molecular Structure
Suppress Noise	Reduce Frequency Level
Remove Heat	Reduce Molecular Movement
Heat Filament	Agitate Molecules
Reduce Friction	Reduce Resistance
Collect Dirt	Collect Elements
Create Image	Change Contrast
Cut Grass	Cut Vegetation
Tighten Bolt	Tighten Hardware
Eat Ham	Eat Animal
Filter Noise	Filter Frequencies
Control Deflection	Control Movement
Educate Students	Increase Intelligence
Prevent Contamination	Remove Impurities
Transport People	Move Creatures
Eat Toast	Consume Starch
Control Anger	Control Mood

Figure 13.3.   Generalizing Functions

Gleaned Functions	Generalized Functions
Rotate Timing Disc	Move Object
Prevent Filament Oxidation	Remove Air
Share Financial Matters	Divide Substance
Share Tactical Intelligence	Distribute Knowledge
Prevent Vibration Disengagements	Inhibit Separations

Figure 13.4.   Functions Generalized

in the earlier chapters. Figure 13.4 also shows how these five functions might be generalized to open new avenues for creativity. After a function has been generalized, you brainstorm the function by inserting it into the How Else Question. Basically, you repeat brainstorming Step 11 by inserting the generalized functions into the How Else Question.

Let's brainstorm the function "distribute knowledge." This time, instead of writing down functions as we answer the question, we will do what is usually done in a brainstorming session. We divorce ourselves from the project and list

Phone	Frequency Modulated Waves
Television	Amplitude Modulated Waves
Radio	Internet
Lasers	Gossip
Fiber Optics	Leaks
Satellites	Informants
Books	Secret Agents
Humans	Newspapers
Pigeons	Magazines
Digital Signals	Lectures

Figure 13.5. Brainstormed Ideas

all the ways we can think of to "distribute knowledge." Figure 13.5 shows the list that was generated for the communication device in Chapter 12. Once the list has been generated, then we can see how each item can be modified so it will perform the function for our project. Since this function originated from the FAST Diagram developed in Chapter 12, which was a communication device for the military, we are unable to explore the potential of any of these ideas.

MOTIVATE PEOPLE VERSUS PROMPT ACTION

Step 13. Develop Undisclosed Functions[1]

This step may seem like a contradiction because the general definition of a function states that the name of a function tells what needs to be accomplished without disclosing the method of accomplishment. However, when you ask the How Question of a why function, the function you name as the answer describes very specifically the method of accomplishment. This fulfills the second definition of a function presented in the Glossary of FAST Terms in Appendix B, which reads as follows:

A function describes a specific method of how to accomplish its next higher level function.

As soon as you disclose the method of doing something, other team members immediately have a tendency to go along with the same line of thinking. On the other hand, if you can creatively write down a function that does not disclose the method recorded in the logic diagram, people's minds take them someplace else. Then when they answer the How Logic Question for that new function, they become more creative. This step requires some deep thinking. The best way to implement this step is to have all participants look at the various functions in a logic diagram and see how many of the primary path functions they can give new names; however, if role-playing and writing your answers as functions is easier, use that method instead. Then those functions are listed, and all participants brainstorm them one by one, the same way we did in Step 12.

Generally, this technique is used for those functions that appear in the higher levels of a FAST Tree or FAST Diagram. Any function may be selected when using this approach. Often a project becomes too large for one team to analyze an entire logic diagram so several teams may be working jointly on the same project, and each team is assigned one of the branches of the tree. The first function in a given branch is the first function a team usually selects for this step. The higher level functions always produce the best creative opportunities. This approach tends to expand and broaden people's understanding of the concept and opens new opportunities for developing new ideas. Perhaps the best way to explain this step is to show some examples. In doing this, we will first limit ourselves to just two levels of some function clusters previously developed in other chapters.

When function clusters are separated horizontally by only one tab space, they are called two-level function clusters even when they contain more than two functions, as demonstrated in the two-level cluster below. The functions in this cluster are taken from FAST Tree 2 in Figure 10.5 for Project 3 in Chapter 10.

> Foster Better Relationships
> Treat People Respectfully
> Remove Tension
> Motivate People

When we initially asked the question

How is *foster better relationships* actually accomplished?

we ended up with the three second-level functions: "treat people respectfully," "remove tension," and "motivate people."

Let's suppose the three people who gave these answers were not allowed to disclose the methods they were thinking of at the time. What functions could they have given instead? This may require considerable thought on their part. What if they had recorded the following three functions for the second level?

> Foster Better Relationships
> Be Courteous
> Transfer Strain
> Prompt Action

EXPANDING NEW FUNCTIONS

When we put these three new functions into the How or How Else Logic Questions and at the same time do a little role-playing and brainstorming, our FAST Tree might have started out looking like the FAST Tree shown in Figure 13.6.

Three new obvious points of view are disclosed and perhaps more if we analyze the results closely. Suppose we merge what we have discovered from this exercise with the completed tree in Figure 10.8 developed in Chapter 10 to analyze love. As we do this, we allow our creativity to add a few other new functions, as shown in Figure 13.7. All new functions formulated by performing this step are identified by placing a pound (#) symbol near the left margin. Notice how quickly the mind starts to work as new points of view come to light.

> Foster Better Relationships
> Be Courteous
> Be Respectful
> Wait Patiently
> Speak Kindly
> Offer Assistance
> Transfer Strain
> Release Tension
> Add Support
> Reduce Tension
> Prompt Action
> Motivate People
> Give Reasons
> Give Rewards

Figure 13.6. New Functions for Love FAST Tree

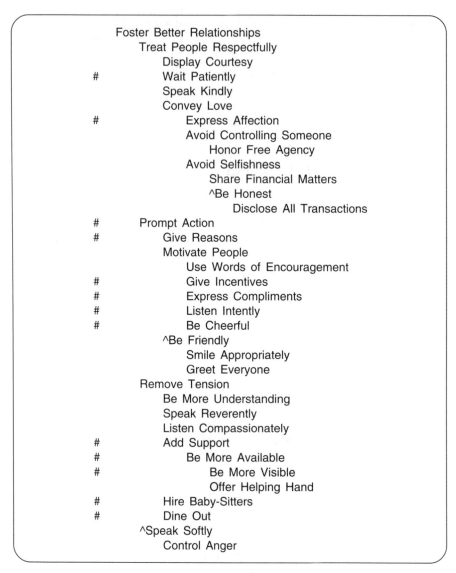

```
              Foster Better Relationships
                 Treat People Respectfully
                    Display Courtesy
        #           Wait Patiently
                    Speak Kindly
                    Convey Love
        #              Express Affection
                       Avoid Controlling Someone
                          Honor Free Agency
                       Avoid Selfishness
                          Share Financial Matters
                          ^Be Honest
                             Disclose All Transactions
        #        Prompt Action
        #           Give Reasons
                    Motivate People
                       Use Words of Encouragement
        #              Give Incentives
        #              Express Compliments
        #              Listen Intently
        #              Be Cheerful
                    ^Be Friendly
                       Smile Appropriately
                       Greet Everyone
                 Remove Tension
                    Be More Understanding
                    Speak Reverently
                    Listen Compassionately
        #           Add Support
        #              Be More Available
        #                 Be More Visible
                          Offer Helping Hand
        #           Hire Baby-Sitters
        #           Dine Out
                 ^Speak Softly
                    Control Anger
```

Figure 13.7. Love FAST Tree 6

We could expand this FAST Tree more, but this is sufficient to demonstrate how deeper thinking can expand your understanding of a project.

Note that this little exercise added 13 new functions to our FAST Tree. Many more come to mind, but this is ample to exhibit the value of performing this step.

We could generalize the function "share financial matters." This could be accomplished by eliminating the modifier or by just changing the function to "share things." After you generalize a function, insert it in the How Else Question, and use your creativity to generate more new functions. An updated FAST Diagram is shown in Figure 13.8. Compare this logic diagram with the FAST Diagram shown in Figure 10.9 in Chapter 10. We could expand this updated diagram even more by adding the 28 functions generated at the end of Chapter 10 during our earlier brainstorming exercise.

ELIMINATE FAILURES VERSUS ERADICATE DEFECTS

Let's consider three more two-level clusters, each taken from the FAST Tree for Project 5 in Chapter 12, which dealt with the military communication device we talked about earlier in this step. The following clusters are gleaned from Figure 12.9:

> Increase Field Communication Performance
> Eliminate Failures

> Eliminate Failures
> Observe Anomaly

> Eliminate Computer Lockups
> Document Lockups

The challenge when performing Step 13 is to name new functions that do not disclose the methods described within the names given to any how functions. In order to do that, we change the three second-level functions listed above into three new functions that lead us down new paths. The initial functions are used to convert or formulate new functions.

Initial Functions		**New Formulated Functions**
Eliminate Failures	into	Eradicate Defects
Observe Anomaly	into	Detect Unusual Behavior
Document Lockups	into	Submit Evidence

We create these new functions by trying to convey to someone else functions that will not disclose what we have already recorded. This requires a bit of creativity in and of itself. A thesaurus is sometimes helpful in formulating these new functions.

How →   ← Why

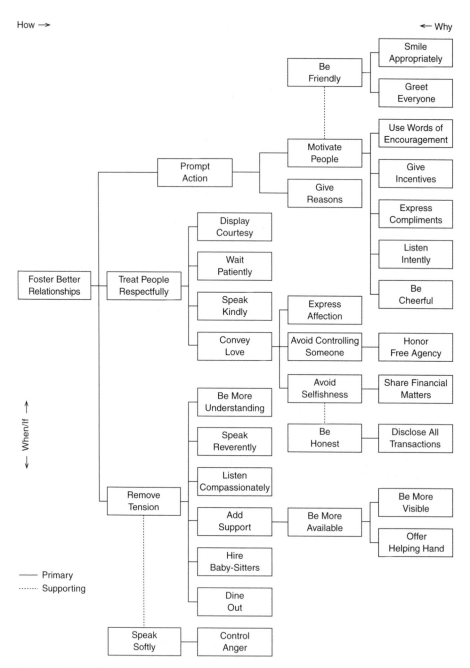

Figure 13.8. **Updated Love FAST Tree**

BRAINSTORMING NEW FORMULATED FUNCTIONS

Let's see what these new functions might do for us if we apply the How Else Logic Question to each one of them and at the same time role-play.

Q   How else can *eradicate defects* be accomplished or performed?

Role   Die caster
A1   Eliminate impurities

Role   Doctor
A2   Eat properly

Role   Contractor
A3   Hire experts

Role   Proofreader
A4   Duplicate effort

This little exercise produced four more new functions; however, only two of them seem to apply to our communication device: "hire experts" and "duplicate effort." This can be changed to "duplicate tests."

We will now apply the How Else Logic Question to the second new function:

Q   How else can *detect unusual behavior* be accomplished or performed?

Role   Observer
A1   Monitor or videotape operators

Role   Therapist
A2   Interview associates

Role   Military officer
A3   Interview those receiving the information

These three answers could be used by a team assigned to find out what operators do when they send, receive, turn on, shut down, and operate the communication equipment discussed in Chapter 12.

Now let's consider the third new function:

Q   How else can *submit evidence* be accomplished or performed?

Role Analyst
A1 Record self-test dates and record LRU replacement dates
A2 Review transmittals

Role Detective
A3 Interview witnesses

Four new functions are listed as answers to this last question. They suggest that we analyze operator performance. We *will not* merge these 11 new functions into the FAST Tree for "support mission" in Project 5 here. That task will be left up to you.

Let's move on and consider four more how functions. Recall that a how function is always a function that is one level lower than a why function. The four second-level functions listed below are gleaned from a nonhardware project that is not included in this book but demonstrates how versatile this last step is. These four second-level functions were obtained from the following two different clusters:

> Teach Principles of Happiness
> > Define Happiness
> > Teach Respect for Human Life
> > Teach Principles of Success

> Obtain Happiness
> > Develop Wholesome Families

These four second-level how functions can be converted as follows:

Initial Functions		**New Formulated Functions**
Define Happiness	into	Felicitate Environment
Teach Respect for Human Life	into	Express Value of Liveliness
Teach Principles of Success	into	Develop Winners
Develop Wholesome Families	into	Create Healthy Household

USING THE THESAURUS

You may be wondering how I was able to come up with these functions which are intended to take us down different thinking paths. Actually, it is fairly easy when using a word-processing program that includes a thesaurus. All you do is sit at a computer and type the function you want to replace, then copy and paste that function several times on separate lines below it, and then replace each verb below the initial function with a different synonym. Do the same thing

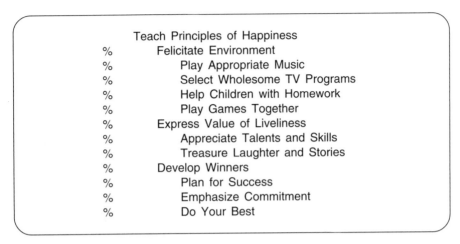

Figure 13.9. New Function Cluster 1

with each noun. Then mentally match up each verb with each noun listed and pick the verb-noun combination that motivates your thinking. You can do the same thing with pencil and paper using *Roget's Thesaurus*. If other words come to mind as you think through this process, consider them also. You can repeat this process on any new formulated functions, thereby increasing your creative opportunities.

As soon as we have several new functions to brainstorm, we create new function clusters by listing these new functions and then brainstorming them to create their second- and third-level functions as well as their higher level functions. We could also insert each new function into the How Else Question and weed out nonpractical ideas as we evaluate our results. We play several different roles as we do this. The results of this activity are displayed in new function cluster 1 in Figure 13.9 and marked with percentage (%) symbols. The new functions conceived for cluster 2 are also displayed in Figure 13.10 using this symbol.

```
              Obtain Happiness
    %         Create Healthy Household
    %            Develop Teamwork
    %            Assign Chores
    %            Contribute Talents and Skills
```

Figure 13.10. New Function Cluster 2

They contribute to our understanding and widen the scope of our project; therefore, they are worthy of consideration and probably should be made a part of the original FAST Tree for "obtain happiness." Consider how long it would take to think of these 16 functions if we did not have this technique to stimulate our thinking.

Let's try this technique one more time on the four new second-level functions we just expanded: "felicitate environment," "express value of liveliness," "develop winners," and "create healthy household." These four functions can be converted in a similar manner as follows:

Initial Functions		**New Formulated Functions**
Felicitate Environment	into	Compliment Surroundings
Express Value of Liveliness	into	Show Appreciation for Life
Develop Winners	into	Produce Leaders
Create Healthy Household	into	Become Domesticated

Once again, we brainstormed these four new formulated functions. After we completed this exercise, we merged our new functions from this activity with the functions we created in Figures 13.9 and 13.10 to form the FAST Tree displayed in Figure 13.11. Note that the functions which appear in Figures 13.9 and 13.10 are still identified with percent (%) symbols, and those resulting from our last brainstorming exercise are identified with ampersands (&). As the earlier information was being merged, additional functions came to mind and were also added and identified using asterisks (*). This logic diagram contains 36 new functions that were not in our earlier diagram. Our final FAST Diagram for "obtain happiness" contains 87 functions, which is too large to display in this book. This diagram and the corresponding Chapter 16 which was used to develop this diagram are available on my web page at www.fastcreativity.com if you are interested in viewing them. I am sure more functions could be added to increase our understanding of this subject.

Perhaps you are wondering how these functions were generated during the brainstorming session. If you think as I do, as I consider the function "compliment surroundings," for example, I imagine myself outside my home or anyplace else and think of what I would like to see. I visualize a nice vegetable garden surrounded by green grass and lots of flowers. Then I think of a function that would be capable of making that become a reality, such as "teach gardening."

Next I do the same type of thinking as I enter a beautiful living room. I visualize myself in that living room and notice how colors and furnishings are coordinated, this time thinking that I could do the same thing if I learned decorating skills. Then I might visualize myself walking into a kitchen or

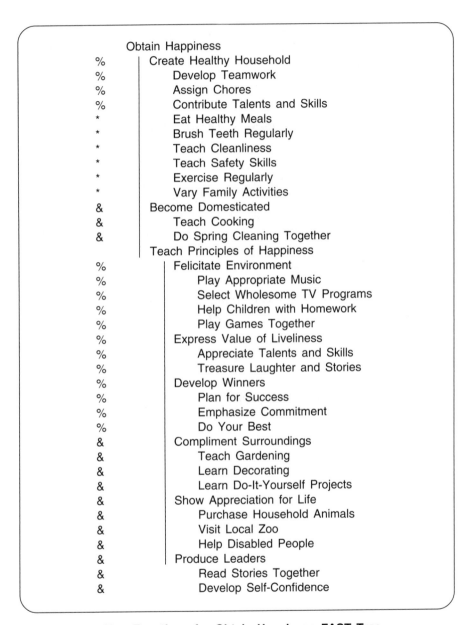

```
            Obtain Happiness
    %       │   Create Healthy Household
    %       │       Develop Teamwork
    %       │       Assign Chores
    %       │       Contribute Talents and Skills
    *       │       Eat Healthy Meals
    *       │       Brush Teeth Regularly
    *       │       Teach Cleanliness
    *       │       Teach Safety Skills
    *       │       Exercise Regularly
    *       │       Vary Family Activities
    &       │   Become Domesticated
    &       │       Teach Cooking
    &       │       Do Spring Cleaning Together
            │   Teach Principles of Happiness
    %       │       Felicitate Environment
    %       │           Play Appropriate Music
    %       │           Select Wholesome TV Programs
    %       │           Help Children with Homework
    %       │           Play Games Together
    %       │       Express Value of Liveliness
    %       │           Appreciate Talents and Skills
    %       │           Treasure Laughter and Stories
    %       │       Develop Winners
    %       │           Plan for Success
    %       │           Emphasize Commitment
    %       │           Do Your Best
    &       │       Compliment Surroundings
    &       │           Teach Gardening
    &       │           Learn Decorating
    &       │           Learn Do-It-Yourself Projects
    &       │       Show Appreciation for Life
    &       │           Purchase Household Animals
    &       │           Visit Local Zoo
    &       │           Help Disabled People
    &       │       Produce Leaders
    &       │           Read Stories Together
    &       │           Develop Self-Confidence
```

Figure 13.11. New Functions for Obtain Happiness FAST Tree

bathroom and notice a faucet dripping. Obviously, my surroundings could be better if the faucet was fixed. I might fix it myself, which sparks the function "learn do-it-yourself projects." Similar role-playing thinking allows other functions to come into a person's mind.

FAST CAN HELP YOU BECOME MORE CREATIVE

FAST is a creative way of analyzing anything that interests you. It is a thinking technique that investigates functions. It forces you to think in a different manner than you may have ever thought before. It is exciting and motivating because you are not always sure where this method of thinking will lead you. It can open new doors for you and stimulate your creativity. It is a systematic method of logically analyzing any subject.

The whole idea behind these last two steps is to expand your thinking into new areas and to see if new functions can be formulated. Only you can determine if these last two steps of this procedure are worthwhile to perform when you analyze a given project.

NOTE

1. Bytheway, C.W., "FAST — An Organized Stimulus to Creativity," SAVE Proceedings, 1975 International Conference, p. 38.

14

OTHER APPLICATIONS OF FAST

FAST can be used in many different ways to increase a person's understanding of a given subject just by writing down a list of functions and then organizing those functions into a FAST Tree. All you need is a pencil and a piece of paper to accomplish the task. The FAST Tree then can be used at a later date to help recall details about the subject and also recall important points quickly. FAST can even help you in your daily activities or in your work environment if you list all the things you do each day as functions and then organize them into a logic diagram.

WHY ANSWERS MOTIVATE PEOPLE

Many other activities might be performed better if you knew why you are actually required to perform them. Some people hate to give talks or teach other people a given subject even though they are very qualified to do so. Perhaps their fears would be eliminated if they put all their thoughts about the subject in writing using functions and then use this technique to organize their thoughts into a logic diagram. As they do this, they will be surprised how much they can expand their subject just by looking at their diagram. A logic diagram becomes an excellent outline for any subject and can be used to ask twice as many questions as there are functions posted in the diagram. This method of presentation can stimulate an audience because it will cause people to be creative as they answer the questions when asked "why" and "how" of any function in the diagram. Questions always tend to increase the level of interest.

A TECHNOLOGICAL ADVANCEMENT

A process or procedure can be easily understood and followed by writing the sequence of operations in chronological order using functions. A product often can be improved or simplified by using the FAST Technique. Writing functions, no doubt, has many other applications that have yet to be discovered. A logic diagram developed by Jerry Kaufman demonstrates the technology advancement of a timepiece.* His FAST model is shown in Figure 14.1.

The following quote indicates other potential uses of FAST:

> On a technical level, Bytheway's contribution was to take the logic associated with the structural paradigm and put it in the form that is intuitive to use. Its primary use is to solve problems, but as a fundamental representation scheme new uses continue to emerge, including a tool for teaching and a way to represent technology for research.
>
> —*Dr. Martin Hyatt, Ph.D.*
> *Developer of Synoptics Creativity*

A book of FAST Diagrams that contain the progressions of all major technological advancements similar to the timepiece could help people understand and appreciate how creativity has made it possible for us to enjoy many great inventions. One other useful application of FAST is writing down functions as you read about a particular subject. Each paragraph of every article you read contains important functions. The procedure for doing this is to formulate a list of functions as you read each paragraph and then organized all of your functions into a logic diagram. Use your creativity as you construct the diagram as new functions come into your mind. Other concepts are frequently thought of during this exercise, so express them also as functions.

A HOME SECURITY CHALLENGE

Try your skill with the following article written by Chris E. McGoey entitled "Home Security: Burglary Prevention Advice." See how many functions you can formulate.

---

* J.J. Kaufman, President, J.J. Kaufman Associates, Inc., Houston, Texas.

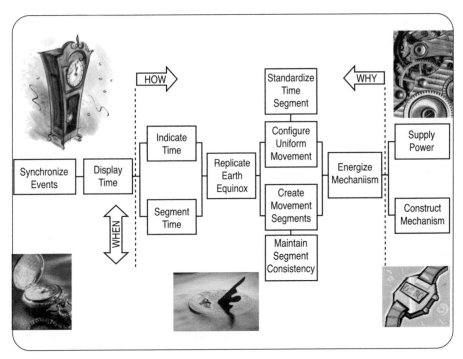

Figure 14.1. Timepiece FAST Model

Home Security: Burglary Prevention Advice*

By Chris E. McGoey, CPP, CSP, CAM

Your home is your castle — or is it? Are you really safe once you get home
and lock your door? In an open society, your home should be the sanctuary for
you and your family. Your home is the only environment where you have
control over who can get close to you or your family. Protecting your home and
family from criminal intrusion should be high on your list of priorities. See my
web site on family security tips for more information on protecting your family
from harm.

Home Burglary

By far, the most common threat to our homes is burglary. According to the FBI,
a burglary occurs somewhere in the United States every 15.4 seconds. By

---

definition, the crime of burglary is a nonconfrontational property crime that occurs when someone is not at home. However, becoming a burglary victim can leave a family feeling vulnerable and violated. To avoid becoming a burglary victim, it is important to first gain an understanding of who commits burglaries and why.

The majority of home and apartment burglaries occur during the daytime, when most people are away at work or school. The summer months of July and August have the most burglaries, and February has the fewest crimes. Burglaries are committed most often by young males under 25 years of age looking for items that are small, expensive, and can easily be converted to cash. Favorite items are cash, jewelry, guns, watches, laptop computers, VCRs, video players, and CDs, and other small electronic devices are high on the list. Quick cash is needed for living expenses and drugs. Statistics tell us that 70 percent of the burglars use some amount of force to enter a dwelling, but their preference is to gain easy access through an open door or window. Ordinary household tools like screwdrivers, channel-lock pliers, small pry bars, and small hammers are most often used by burglars. Burglars continue to flourish because police can only clear about 13 percent of all reported burglaries and rarely catch the thief in the act.

Although home burglaries may seem random in occurrence, they actually involve a selection process. The burglar's selection process is simple. Choose an unoccupied home with easiest access, the greatest amount of cover, and the best escape routes. What follows is a list of suggestions to minimize your risk by making your home unattractive to potential burglars.

Doors and Locks

The first step is to "harden the target" or make your home more difficult to enter. Remember that burglars will simply bypass your home if it requires too much effort or requires more skill and tools than they possess. Most burglars enter via the front, back, or garage door. Experienced burglars know that the garage door is usually the weakest point of entry, followed by the back door. The garage and back doors also provide the most cover. Burglars know to look inside your car for keys and other valuables, so keep it locked, even when parked inside your garage. Use high-quality Grade-1 or Grade-2 locks on exterior doors to resist twisting, prying, and lock-picking attempts. A quality dead bolt lock will have a beveled casing to inhibit the use of channel-lock pliers used to shear off lock cylinder pins. A quality door knob-in-lock set will have a "dead latch" mechanism to prevent slipping the lock with a shim or credit card.

- Use a solid core or metal door for all entrance points
- Use a quality, heavy-duty dead bolt lock with a one-inch throw bolt
- Use a quality, heavy-duty knob-in-lock set with a dead latch mechanism
- Use a heavy-duty, four-screw strike plate with three-inch screws to penetrate into a wooden door frame
- Use a wide-angle 160-degree peephole mounted no higher than 58 inches

The most common way to force entry through a door with a wooden jamb is to simply kick it open. The weakest point is almost always the lock strike plate that holds the latch or lock bolt in place, followed by a glass paneled door. The average door strike plate is secured only by the soft-wood door jamb molding. These lightweight moldings are often tacked on to the door frame and can be torn away with a firm kick. Because of this construction flaw, it makes sense to upgrade to a four-screw, heavy-duty, high-security strike plate. They are available at most quality hardware stores and home improvement centers and are definitely worth the extra expense. Install this heavy-duty strike plate using three-inch wood screws to cut deep into the door frame stud. Use these longer screws in the knob lock strike plate as well, and use at least one long screw in each door hinge. This one step alone will deter or prevent most through-the-door forced entries. You and your family will sleep safer in the future.

Sliding Glass Patio Doors

Sliding glass doors are secured by latches, not locks. They are vulnerable to being forced open from the outside because of these inherently defective latch mechanisms. This easily can be prevented by inserting a wooden dowel or stick into the track, thus preventing or limiting movement. Other blocking devices available are metal fold-down blocking devices called "charley bars" and various track blockers that can be screwed down.

The blocking devices described above solve half of the equation. Older sliding glass doors can be lifted up and off their track and thereby defeat the latch mechanism. To prevent lifting, you need to keep the door rollers in good condition and properly adjusted. You can also install antilift devices, such as a pin that extends through both the sliding and fixed portion of the door. There are also numerous locking and blocking devices available at any good quality hardware store that will prevent a sliding door from being lifted or forced horizontally. Place highly visible decals on the glass door near the latch mechanism which indicates that an alarm system, a dog, or a block watch/operation identification is in place. Burglars dislike alarm systems and definitely big barking dogs.

- Use a secondary blocking device on all sliding glass doors
- Keep the latch mechanism in good condition and properly adjusted
- Keep sliding door rollers in good condition and properly adjusted
- Use antilift devices such as through-the-door pins or upper track screws
- Use highly visible alarm decals, beware-of-dog decals, or block watch decals

Windows

Windows are left unlocked and open at a much higher rate than doors. An open window, visible from the street or alley, may be the sole reason for your home to be selected by a burglar. Ground floor windows are more susceptible to break-ins for obvious reasons. Upper floor windows become attractive if they can be accessed from a stairway, tree, fence, or balcony. Windows have latches, not locks, and therefore should have secondary blocking devices to prevent sliding them open from the outside. Inexpensive dowels and sticks work well for horizontal sliding windows, and through-the-frame pins work well for vertical sliding windows. For ventilation, block the window open no more than six inches and make sure someone cannot reach in from the outside and remove the blocking device or reach through and unlock the door.

In sleeping rooms, these window-blocking devices should be capable of being removed easily from the inside to comply with fire codes. Like sliding glass doors, antilift devices are necessary for ground-level and accessible aluminum windows that slide horizontally. The least expensive and easiest method is to install screws halfway into the upper track of the moveable glass panel to prevent it from being lifted out in the closed position. As a deterrent, place highly visible decals on the glass door near the latch mechanism which indicate that an alarm system, a dog, or a block watch/operation identification system is in place.

- Secure all accessible windows with secondary blocking devices
- Block accessible windows open no more than six inches for ventilation
- Make sure someone cannot reach through an open window and unlock the door
- Make sure someone cannot reach inside the window and remove the blocking device
- Use antilift devices to prevent windows from being lifted out
- Use crime prevention or alarm decals on ground-accessible windows

Be a Good Neighbor

Good neighbors should look out for each other. Get to know your neighbors on each side of your home and the three directly across the street. Invite them into your home, communicate often, and establish trust. Good neighbors will watch out for your home or apartment when you are away, if you ask them. They can report suspicious activity to the police or to you while you are away. Among them, good neighbors can see to it that normal services continue in your absence by allowing vendors to mow your lawn or remove snow. Good neighbors can pick up your mail, newspapers, and handbills and can inspect the outside or inside of your home periodically to see that all is well. Good neighbors will occasionally park in your driveway to give the appearance of occupancy while you are on vacation.

Allowing a neighbor to have a key solves the problem of hiding a key outside the door. Experienced burglars know to look for hidden keys in planter boxes under doormats and above the ledge. Requiring a service vendor to see your neighbor to retrieve and return your house key will send the message that someone is watching. This neighborhood watch technique sets up what is called "territoriality," which means that your neighbors will take ownership and responsibility for what occurs in your mini-neighborhood. This concept works in single-family homes, communities, and apartment properties. This practice helps deter burglaries and other crimes in a big way. Of course, for this to work, you must reciprocate and offer the same services.

- Get to know all your adjacent neighbors
- Invite them into your home and establish trust
- Agree to watch out for each other's home
- Do small tasks for each other to improve territoriality
- While on vacation, pick up newspapers and flyers
- Offer to park your car in your neighbor's driveway
- Return the favor and communicate often

Lighting

Interior lighting is necessary to show signs of life and activity inside a residence at night. A darkened home night after night sends the message to burglars that you are away on a trip. Light timers are inexpensive and can be found everywhere. They should be used on a daily basis, not just when you are away. In this way, you set up a routine that your neighbors can observe, which will allow them to become suspicious when your normally lighted home becomes dark. Typically, you want to use light timers near the front and back windows, with the curtains drawn. The pattern of the light dimmers clicking on and off should simulate actual occupancy. It is also comforting not to have to enter a dark residence. The same light timers can be used to turn on radios or television sets to further enhance the illusion of occupancy.

Exterior lighting is also very important. It becomes critical if you must park in a common area parking lot or underground garage and need to walk to your front door. The purpose of good lighting is to allow you to see if a threat or suspicious person is lurking in your path. If you can see a potential threat in advance, then you at least have the choice to avoid it. Exterior lighting needs to be bright enough for you to see 100 feet, and it helps if you can identify colors. Good lighting is definitely a deterrent to criminals because they do not want to be seen or identified.

Another important area to be well lighted is the perimeter of your home or apartment, especially at the entryway. Exterior lighting on the front of a property should always be on a timer to establish a routine and the appearance of occupancy at all times. Common area lighting on apartment properties also should be on a timer or photocell to turn on at dusk and turn off at dawn. The practice of leaving the garage or porch light turned on all day at a single-family home is a dead giveaway that you are out of town. Exterior lighting at the rear of a home or apartment usually is on a switch because of the proximity to the sleeping rooms. The resident can choose to leave these lights on or off. Security lights with infra-red motion sensors are relatively inexpensive and easily can replace an exterior porch light or side door light on a single-family home. The heat-motion sensor can be adjusted to detect body heat and can be programmed to reset after one minute. These security lights are highly recommended for single-family homes.

- Use interior light timers to establish a pattern of occupancy
- Exterior lighting should allow 100-foot visibility
- Use good lighting along the pathway and at your door
- Use light timers or photocells to turn lights on/off automatically
- Use infra-red motion sensor lights on the rear of single-family homes

Alarm Systems

Alarm systems definitely have a place in a home security plan and are effective, if used properly. The reason why alarm systems deter burglaries is because they increase the potential fear of being caught and arrested by the police. The deterrent value comes from the alarm company sign and from the alarm decals on the windows. Home and apartment burglars usually will bypass a property with visible alarm signs and go to another property without such signs. Some people with alarm systems feel that these signs and decals are unsightly and will not display them. The risk here is that an uninformed burglar might break a window or door and grab a few quick items before the police can respond. Also, do not write your alarm pass code on or near the alarm keypad.

Alarm systems need to be properly installed and maintained. Alarms systems can monitor for fire as well as burglary for the same price. All systems should have an audible horn or bell to be effective in case someone does break in. However, these audible alarms should be programmed to reset automatically after one or two minutes. The criminal got the message and will be long gone, but neighbors will have to listen to the alarm bell, sometimes for hours, until it is shut off. If you use a central station to monitor your alarm, make sure your response call list is up to date. Home alarms, like car alarms, generally are ignored except for a brief glance. However, if you have established and nurtured your neighborhood watch buddy system, you will experience a genuine concern by your neighbor. It is not unusual to have a neighbor wait for the police, allow them inside for an inspection, and secure the residence. A good neighbor can also call the glass company or locksmith to repair any damage, if preauthorized by you.

The greatest barrier in getting to this level of neighborhood participation is taking the first step. You can get help by calling your local crime prevention unit at the police department. Most police departments in large cities have neighborhood watch coordinators to help you set this up. You should invite your adjacent neighbors over to your home for coffee and begin the information exchange. You'll be amazed how the process runs on automatic from there.

- Alarm systems with visible signage are effective deterrents
- Alarm systems need to be properly installed, programmed, and maintained
- Alarm systems need to have an audible horn or bell to be effective
- Make sure your alarm response call list is up to date
- Instruct your neighbors how to respond to an alarm bell

Home Safes

Since the price of good home safes is falling, having a safe in your home is a wise investment. Home safes are designed to keep the smash-and-grab burglar, nosey kids, dishonest baby-sitter, or housekeeper from gaining access to important documents and personal property. Home safes need to be anchored into the floor or permanent shelving.

- Use the safe every day so it becomes routine
- Protect the safe code and change it occasionally
- Install the safe away from the master bedroom or closet

Operation Identification

This is a program supported by most police agencies. They recommend that you engrave your driver's license number, not your social security number, on televisions, stereos, computers, and small electronic appliances. They suggest this so they can identify and locate you if your stolen items are recovered. I suggest that you go way beyond this step.

I recommend that you photograph your valuables in their locations around your home and make a list of the make, model, and serial number. This is very important for proof when filing insurance claims. You should keep this list in a safety deposit box or with a relative for safekeeping. Keep receipts of the larger items in case you need to prove the value of the items for insurance purposes. Beyond that, I recommend that you photocopy important documents and the contents of your wallet. You will be thankful that you took these steps in case your home is ever destroyed by fire or flood, is ransacked, or if your wallet is lost or stolen.

- Identify your valuables by engraving your driver's license number
- Photograph and record the serial number of all valuables
- Photocopy the contents of your wallet and other documents
- Store the copies in a safety deposit box or with a relative

Logic Diagram

I'm sure you found this article very useful, but can you remember everything Mr. McGoey told you to do in order to keep your home safe? You now have the opportunity to develop a logic diagram to aid you in recalling the whys and hows of his article. Take a few minutes and organize the functions you compiled while reading this article. When you go to a hardware store to buy a lock for your front door, will you purchase the kind of lock he suggested? A logic diagram will help you remember the important facts. Compare your results with the FAST Tree in Figure 14.2. If you find other applications for this technique, I would love to hear from you at www.fastcreativity.com.

Note that the completed FAST Tree contains basically the same information written in the article. If you want to know how to implement a particular branch of the tree, just go to the function you are interested in and perform the functions indented below that function.

```
Secure Residence
    Prevent Intruder Entry
      | Deter Intruders
      |       Post Lawn Security/Alarm Signs
      |       Post Door/Window Security/Alarm Signs
      |       Install Sensor Lights
      |             Install Infra-red Motion Sensors
      |       Use On/Off Light Timers
      |       Close Curtains/Drapes
      |       Install Door/Window Metal Bars
      |       ^Ensure Emergency Exit
      | Deter Vacation Intruders
      |       Request Neighbor Assistance
      |       ^Identify Assistance Desired
      |             Remove Snow
      |             Remove Newspapers/Handbills
      |             Pick Up Mail
      |             Suggest Driveway Parking
      |             Report Suspicious Activity
      |             Respond to Alarm
      |       ^Authorize Emergency Action
      |       Use On/Off TV/Radio Timers
      | Secure Outside Doors
      |       Secure Entrance Doors
      |             Install Solid Door Jambs
      |             Install Solid Core/Metal Door
      |                   Purchase Doors
      |                   Install Hinges
      |                         Use Standard Screws
      |                         Use One Three-Inch Screw per Hinge
      |                   Install Heavy-Duty Knob-in-Lock Set
      |                         Purchase Grade-1 or Grade-2 Locks
      |                         Purchase Dead-Latch Mechanism
      |                         Use Three-Inch Screws in Knob Strike Plate
      |                         Purchase One-Inch Throw Bolt
      |                         Purchase Tapered Casing
      |                   ^Supply Key
      |                         Select Trusted Neighbor
      |                   Install Heavy-Duty Lock Strike Plate
      |                         Use Three-Inch Screws
      |             Install Heavy-Duty Dead Bolt
      |             Install 160-Degree Peephole
      |                   Position <58 Inches High
```

Figure 14.2.   Secure Residence FAST Tree

^Avoid Glass Paneled Doors
Secure Sliding Doors
    Prevent Movement
        Use Horizontal/Vertical Blockers
    ^Post Visible Alarm Decals
Secure Windows
    Close/Lock Windows
    Prevent Movement
        Use Screw in Upper Track
        Use Window Blockers
    ^Post Visible Alarm Decals
Secure Garage Doors
    Prevent Vertical Movement
        Disengage Opener Power (Vacation Safety)
        Use Vertical Blocker
^Secure Stored Vehicles
    Lock All Vehicles
Install Alarm System
^Avoid Writing Pass/Entry Codes
Secure Valuables
    Install Home Safe
    Anchor Safe
^Ensure Property Repossession
    Engrave Driver's License Number (on Electronic Items)
    Photograph Valuables
    Make List (Make, Model, Serial Number of Items)
    Keep Major Purchase Receipts
    Photo Wallet Contents

Legend:
Why functions are posted one tab space to the left and above the function being investigated.

How functions are posted one tab space to the right and below the function being investigated.

Supporting functions are posted directly below the function they support preceded by a caret (^) symbol.

The *basic function* to be performed is the top function in this FAST Tree.

Figure 14.2 continued.

15

SUMMARY OF
FAST PROCEDURE

A brief description of the 13 steps of the FAST Creativity procedure is provided in this chapter without any project details to distract from the process. If a particular step needs clarification, search for examples in Chapters 8 through 13. Each step is easy to locate in those chapters as a heading identifies where each step starts.

STEP 1. SELECTING A PROJECT

Step 1 of the procedure is to select a subject or project to analyze. The five methods for doing this are listed in Figure 15.1.

The questions for Methods 3, 4, and 5 are shown in Figures 15.2, 15.3, and 15.4, respectively.

STEP 2. SELECTING PARTICIPANTS

No doubt, you will always want someone else to assist you in developing a logic diagram because different points of view and discussions stimulate creativity. Make sure the people you ask understand the basic concepts of FAST Creativity. Try to develop a couple of diagrams by yourself before asking other people to assist you. Start by constructing a FAST Diagram as described in Chapter 4 and in Chapters 8 through 12 by writing functions on small cards and arranging them until the logic holds in both directions. Participants you select should

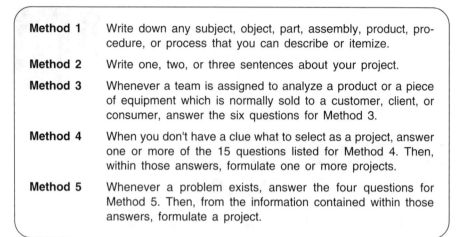

Method 1   Write down any subject, object, part, assembly, product, procedure, or process that you can describe or itemize.

Method 2   Write one, two, or three sentences about your project.

Method 3   Whenever a team is assigned to analyze a product or a piece of equipment which is normally sold to a customer, client, or consumer, answer the six questions for Method 3.

Method 4   When you don't have a clue what to select as a project, answer one or more of the 15 questions listed for Method 4. Then, within those answers, formulate one or more projects.

Method 5   Whenever a problem exists, answer the four questions for Method 5. Then, from the information contained within those answers, formulate a project.

Figure 15.1. Methods for Selecting Projects

Q1   What product or piece of equipment have you been assigned to analyze?

Q2   What is the main purpose for which this product has been built or assembled?

Q3   How can this product be made so it is more dependable?

Q4   How can this product be made so it is more convenient to use?

Q5   How can this product be improved?

Q6   How can this product be made so it is more pleasing to the five senses?

Figure 15.2. Questions for Method 3

be willing to follow the procedure. People with different skills and backgrounds generally provide a good mix.

A typical team for constructing a FAST Diagram for a *consumer-type product* consists of five decision makers from within the manufacturing organization. These participants are usually supplied from the following departments:

- Engineering
- Manufacturing engineering
- Purchasing

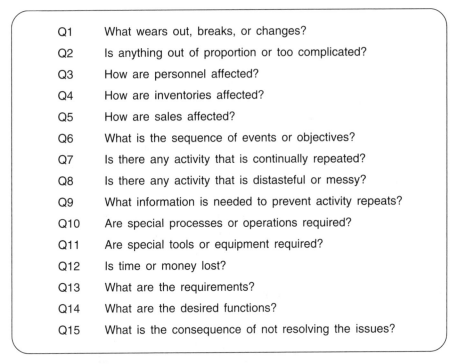

Q1	What wears out, breaks, or changes?
Q2	Is anything out of proportion or too complicated?
Q3	How are personnel affected?
Q4	How are inventories affected?
Q5	How are sales affected?
Q6	What is the sequence of events or objectives?
Q7	Is there any activity that is continually repeated?
Q8	Is there any activity that is distasteful or messy?
Q9	What information is needed to prevent activity repeats?
Q10	Are special processes or operations required?
Q11	Are special tools or equipment required?
Q12	Is time or money lost?
Q13	What are the requirements?
Q14	What are the desired functions?
Q15	What is the consequence of not resolving the issues?

Figure 15.3. Questions for Method 4

Q1	What problem shall we discuss?
Q2	Why do you think this is a problem?
Q3	Why do you think a solution is needed?
Q4	What is there about this problem area that disturbs you?

Figure 15.4. Questions for Method 5

- Estimating
- Marketing

When analyzing *manufacturing procedures,* consider using:

- The person responsible for the manufacturing process
- A worker who is capable of performing almost any assignment

- An industrial engineer
- A foreman
- A product engineer for a high-volume product

Other areas to be analyzed such as schools; local, state, and federal governmental agencies and services; and religious and community projects all require judgment in selecting participants. You should always try to involve the people who will be affected by any proposed changes. For example, a school project might involve a teacher, counselor, PTA member, police officer, student, and a citizen. The best way to obtain acceptance is to involve the people who will be affected in the logic thinking exercise. Other approaches are covered in Chapter 6. Chapter 6 also goes into some detail describing how to develop Step-by-Step Diagrams and Composite Diagrams, which are usually constructed using the Internet with participants located at various places throughout the world.

STEP 3. INITIAL FUNCTIONS

This step requires the answers, statements, and facts obtained during Step 1 to be transposed into function names. If physical parts or equipment are involved, their names should be supplied along with drawings, if available, so all participants have the same information to start their analysis. Each participant should develop his or her own list. Combine the lists generated by all participants into a single list and massage it until all agree that the functions represent what is being analyzed. The type of logic diagram that is to be created should be determined before proceeding to Step 4. The types are individual, step-by-step, composite merged, and normal.

STEP 4. INITIAL BASIC FUNCTION

The initial basic function is obtained by selecting the function in the list developed during Step 3 that appears to be the most important and inserting it into the following question:

> If *this function* didn't have to be performed, would any of the other functions listed still have to be accomplished or performed?

If a "no" answer is obtained for all the other functions, you have identified the initial basic function. If a "yes" answer is obtained, insert the function that gives you this answer into this same question. See Step 4 in Chapter 8 if you need more direction concerning this step.

STEP 5. DEVELOP HIGHER LEVEL FUNCTIONS

This step requires a diligent search for higher level functions. It is accomplished by inserting the initial basic function in the blanks where the asterisks appear in the following three questions:

1. Why is it necessary to _____*_____?

2. What higher level function caused _____*_____ to come into being?

3. What is really trying to be accomplished when _____*_____ is performed?

Keep in mind that when these questions are asked, we are trying to come up with three new functions; therefore, we have to force ourselves to think more deeply about our subject or project. Glean from the answers obtained from this exercise three or more functions and put them along with the initial basic function in a list named "Higher Level Functions."

STEP 6. IDENTIFYING THE BASIC FUNCTION

From the list of higher level functions developed during Step 5, select the function you think is the most important and insert it into the following Basic Function Determination Question:

If **this function** didn't have to be performed, would any of the other (higher level) functions still have to be performed?

If all other functions yield "no" answers, you have selected the basic function. Indicate this fact by placing an asterisk (*) in front of it in your higher level function list.

STEP 7. DEVELOP PRIMARY PATH FUNCTIONS

Start this step by listing all the functions you have formulated for your project up to this point, and identify the list as "List of Functions." If you are developing a FAST Diagram, write each function on a small card to form this list. Put the cards on a flat surface and start to develop the diagram by placing the basic function card at your left. You should allow room to place additional cards to the right and above and below this card as you develop your logic diagram as you perform the remaining steps of the procedure. The next step is to ask the How Logic Question of the basic function. A FAST Diagram progressively grows from left to right, as demonstrated in Chapter 4.

If you are developing a FAST Tree, post your basic function on a line in a word-processing program and then ask the How Logic Question of it:

> How is *this function* actually accomplished or proposed to be accomplished?

Type your answer as a function on its own line one tab space to the right and below the function you inserted into the question. If you formulate more than one function for your answer to this question, place each additional function directly below the first function at the same tab spacing. Next ask the following Why Logic Question of each second-level new function. Your Why Logic Question answers should all yield the basic function.

> Why must *this function* be performed?

If the answer is different from the basic function, massage your functions until the logic agrees in both directions. By that I mean, you search for functions that will satisfy the logic and replace those that do not. Then ask the How Logic Question of all new functions, and check the answers to see if they are correct logically. Repeat this process until you are unable to conceive of any more functions. As you do this, check to see if any of the functions in your list of functions fit into the primary path. If any have been included, denote this fact by placing a "7" near the left margin to indicate that a function has been merged during this step. If in doubt about how to accomplish this, refer to Chapter 8.

STEP 8. EVALUATE REMAINING FORMULATED FUNCTIONS

As you start this step, inspect your list of functions created at the beginning of Step 7 and copy all functions that were not merged and paste them at the end of your working file. Tab them in about three tab spaces, and separate each function with at least three lines. It is a good idea to highlight each function in bold before proceeding, so you know each function you started with. Then develop a mini-FAST Tree for each function by asking the Why-How Logic Questions of the functions you just pasted. These are called "function clusters," which consist of at least three functions but may include as many as five or six functions. If in doubt about how to do this, refer to Step 8 for Project 1 in Chapter 8.

Once these clusters have been developed, they are checked to see if the logic is correct. Since the highlighted functions were not included in the primary path, some of them may be supporting functions, and some of the new functions you added during this step may also be supporting functions. Therefore, check all of the how functions of each cluster to see if any support its why function. A how function is always one logic level below a why function in a FAST Tree and is always the next function to the right in a FAST Diagram. The following Verification Question is used to check to see if the logic is correct:

Does *this how function* help *its why function*?

If the answer is yes, then the logic is correct. If the answer is no, then the how function is a supporting function. A caret (^) symbol is placed in front of each how function that yields a "no" to indicate this fact. Since more than one how function may be required to describe how to perform or accomplish a given function, every how function should be checked. Note that a how function may only perform part of the task that must be performed. This is why help is used in this question. Sometimes participants cannot agree on the answer. When this happens, the second Verification Question is asked:

If *this how function* is accomplished, will it help *its why function*?

After the verification has been completed, a copy of the FAST Tree developed during Step 7 is copied and then pasted at the end of the working file, and the pasted tree may be renumbered. The renumbering of the tree is only for reference purposes and is not necessary, since the working file is normally only for a given project. The primary path of the tree usually contains one of the higher level functions of each cluster. When and if it does, those functions are then added to the FAST Tree, and a "7" is typed near the margin of that function and an "8" is typed next to all lower level functions of that cluster. Supporting functions that exist in the clusters and their lower level functions are not merged until Steps 9 and 10.

These function clusters also can be constructed by writing functions on long strips of card stock and arranging them in the same manner as in a computer file, or the functions can be written on small rectangular card stock and arranged like a miniature FAST Diagram. When the higher level functions match the primary path, the cards are then physically moved and made part of the logic diagram under construction. Chalk or dry erase markers can be used on a display board or flip chart to organize these function clusters. Also, a pencil or pen and paper can be used to accomplish the same thing.

STEP 9. USE WHEN/IF LOGIC TO ADD SUPPORTING FUNCTIONS

Next, investigate each of the functions in your logic diagram developed thus far by inserting each primary path function into the following question:

> When/if *this function* is performed, what other functions must be performed?

Add any supporting functions already identified during Step 8, if at all possible, and try to add some new functions if you can. Just the supporting functions are merged during this step by placing a caret (^) symbol and the function name directly below the function being supported. When a how function in a function cluster is a supporting function, it is posted in a FAST Tree directly below the function it supports instead of one tab space to the right as is the case in a cluster. Remember to copy and paste the FAST Tree at the end of your file before you add supporting functions. This permits you to always have a record of what was accomplished during each step of this procedure. Then place a "9" near the margin of each supporting function merged.

STEP 10. DEVELOP SECONDARY PATH FUNCTIONS

Copy your latest logic diagram and paste it at the bottom of your working file. Then use the How Logic and then the Why-How Logic to develop your secondary path functions for each of the supporting functions. Place a "10" near the left margin in line with each function merged during this step within each function cluster. Any function clusters not yet merged during this exercise should be copied and pasted at the end of the working file. These remaining clusters should be expanded by asking "why" of the highest level function until you are able to formulate a function that allows each cluster to be merged. If a cluster is out of the scope of your project, cross it out. Always check the logic of your completed diagram.

STEP 11. BRAINSTORMING HIGHER LEVEL FUNCTIONS

Once your logic diagram has been completed, select three or four of the higher level functions and insert them into the following How Else Question. Follow the ground rules for brainstorming shown in Figure 15.5.

1. Every idea or suggestion is to be recorded.
2. No criticizing of anyone else's ideas or suggestion is allowed. As a reminder of this rule, I like to have everyone write "I will not criticize!" on a full sheet of paper and then crumple the paper and put it in their pocket or keep it in one hand while the brainstorming session is taking place.
3. Divorce yourself from the present project.
4. Play different roles as you think of the function.
5. Ignore what you have experienced or learned in school.
6. Present silly or ridiculous ways of doing things.
7. Disregard standards and traditions.
8. Hitchhiking on other ideas is allowed.
9. Improvements to ideas presented are encouraged.
10. Omit anyone from your group who might be intimidating, if possible.
11. Consider how physical and life sciences would perform the function.
12. Consider primitive and mass-production methods.

Figure 15.5.   Ground Rules for Brainstorming

How else can *this function* be accomplished or performed?

STEP 12. GENERALIZING FUNCTIONS

Perform this step by scanning the functions in your FAST Tree or FAST Diagram and see if you can generalize any of the important functions. Examples of functions generalized appear in Figure 15.6. Insert your new generalized functions into the How Else Question to expand your understanding and stimulate your creativity.

STEP 13. DEVELOP UNDISCLOSED FUNCTIONS

This step pertains to the how functions in your completed logic diagram, which disclose how you intend to perform several why functions. It requires you to formulate new functions that do not disclose the methods described by the names of existing how functions. Examples of functions that might appear in your logic diagram are shown in the left column of Figure 15.7. The right

Initial Functions	Generalized Functions
Attract Attention	Stimulate Senses
Convert Energy	Change Molecular Structure
Educate Students	Increase Intelligence
Cut Grass	Cut Vegetation
Heal Filament	Agitate Molecules
Eat Toast	Consume Starch
Control Anger	Control Mood
Catch Snake	Capture Animal

Figure 15.6. Examples of Generalized Functions

column lists the corresponding names of new functions you might formulate to replace them. The purpose of performing this step is to create new functions which force your mind down different paths that expand your understanding of your project and stimulate additional creativity. The final creative act of this procedure is accomplished by repeating brainstorming in Step 11 by inserting any new functions you formulated into the How Else Question, as was demonstrated in Chapter 13.

PROCEDURE FAST TREE

The Procedure FAST Tree shown in Figure 15.8 is to aid you in following the outlined procedure. Since the steps of this procedure should be followed in sequential order, the FAST Tree is constructed so that all one has to do is start at the top and perform each task as he or she goes down the tree. Steps in the procedure are numbered so you can quickly go to that section of this chapter for more detailed information. This is an example of a logic diagram that is better displayed as a FAST Tree than as a FAST Diagram.

Change This Function	To This Undisclosing Function
Define Happiness	Felicitate Environment
Felicitate Environment	Compliment Surroundings
Motivate People	Prompt Action
Observe Anomaly	Detect Unusual Behavior
Document Lockups	Submit Evidence
Create Healthy Household	Become Domesticated

Figure 15.7. Examples of Undisclosing Method

Follow Procedure
1. Select Project
2. Select Participants
3. Identify Initial Functions
   Supply Names/Drawings
   Share Initial Functions
   Determine Diagram Type
4. Determine Initial Basic Function
5. Develop Higher Level Functions
   List Higher Level Functions
6. Determine Basic Function
7. Develop Primary Path Functions
   Make Function List
   Post Basic Function
   Apply How Logic
   Apply Why-How Logic
   Verify Tree Logic
8. Evaluate Remaining Functions
   Develop Function Clusters
   Identify Supporting Functions
   Merge Primary Path Clusters
9. Identify/Add Supporting Functions
   Use When/If Logic
   Merge Supporting Functions
10. Develop Secondary Path Functions
    Apply How Logic
    Apply Why-How Logic
    Merge Supporting Function Clusters
    Investigate Remaining Functions
    Apply Why Logic
    Merge/Cross Out Clusters
11. Brainstorm Higher Level Functions
12. Generalize Functions
    Brainstorm Generalized Functions
13. Develop Undisclosed Functions
    Brainstorm Undisclosed Functions

Figure 15.8.   Procedure FAST Tree

APPENDIX A: CONSTRUCTING FAST DIAGRAMS

CREATE PROFESSIONAL FAST DIAGRAMS

This chapter discusses how to quickly create professional-looking FAST Diagrams using Microsoft Visio® and by utilizing a number of templates I created using that program to convert FAST Trees into FAST Diagrams. The best approach is to lay out a rough draft of your proposed FAST Diagram on a large sheet of paper, flip chart, flat surface, or dry eraser board by writing functions on rectangular card stock or writing functions within drawn rectangles and placing them so they can be logically connected to show relationships among the various functions. Then connect the rectangles containing functions with horizontal and vertical lines so all logical relationships can be tied together. Those relationships are easy to establish by following the logic of the functions posted within your FAST Tree. Once this has been accomplished, count the maximum number of functions that must be placed from left to right in your rough-draft FAST Diagram. Also, count the maximum number of functions required from top to bottom or vertically.

TEMPLATES AVAILABLE

Once the maximum horizontal and vertical numbers have been determined, decide if the diagram should be drawn portrait or landscape. Available portrait

3x9Template.vst
3x13Template.vst
5x14Template.vst
5x18Template.vst
6x13Template.vst
9x14Template.vst
11x15Template.vst

Figure A.1.   8.5 × 11 Portrait Template Files

templates are listed in Figure A.1, and landscape templates are listed in Figure A.2. The first number of the template name identifies the number of horizontal rectangles and the second number identifies the number of vertical rectangles. Select a template name that is equal to or larger than what is required.

After selecting the appropriate template, start by typing the function names into those rectangles that best match your rough draft. Each template will have more rectangles than is required to create your FAST Diagram. Once all functions have been typed within a rectangle, delete all empty rectangles. Next, connect horizontal, vertical, and other required lines to the functions using the features of Microsoft Visio®. Text blocks can be used to add titles and other information as necessary. It is a good idea to add the Why-How Logic notation and arrows. The templates already include these symbols.

VISIT THE WEB

The Visio® templates listed in Figures A.1 and A.2, the thought-provoking questions, the Procedure FAST Tree, and brainstorming rules are available from the Web Added Value™ Download Resource Center at www.jrosspub.com.

3x10Template.vst
4x10Template.vst
5x10Template.vst
6x10Template.vst
8x10Template.vst
12x10Template.vst
15x11Template.vst

Figure A.2.   11 × 8.5 Landscape Template Files

Visit my Web page at www.fastcreativity.com if you want professionally drawn FAST Diagrams created for you. Portrait, landscape, tabloid, and larger size drawings are available for a fee. All that must be supplied is a copy of your FAST Tree, which can be transmitted over the Internet. Other items of interest are also available on my Web page.

APPENDIX B:
GLOSSARY OF
FAST TERMS AND
THOUGHT-PROVOKING
QUESTIONS

GLOSSARY OF FAST TERMS

Basic Function — A function that describes the principal task that must be performed when analyzing a subject or project. The basic function of a product represents the most significant purpose for which it exists.

Basic Function Determination Technique — The logic and reasoning process used by an analyst when determining the dependence or independence of a list of functions with respect to a given function.

Brainstorming — A creative technique developed by Alex Osborn. A group of participants select a subject and then generate ideas about that subject without engaging in any criticism. The FAST procedure utilizes this technique by inserting several high logic level functions into the How Else Question to creatively find alternate concepts or solutions (see Steps 11, 12, and 13).

Creative Hitchhiking — The process by which new ideas are conceived when someone else's idea is changed or modified or is used to stimulate a person's creative powers.

Dependent Function — Any function that depends upon a higher level function for its existence. In other words, whenever a method is selected for performing a higher level function, the functions below it are brought into existence.

FAST — The acronym for Function Analysis System Technique. This technique, developed by the author, is used to analyze the functions of any subject or project by determining why and how those functions are performed and establishes their relationships to each other. As the technique is applied, communication is increased, analysis is enhanced, and opportunities for creativity become readily available.

FAST Diagram — A diagram that displays all functions that pertain to a given subject or project in an organized manner so that the *cause and effect relationship among the functions is maintained.* Higher level functions appear in the left portion of the diagram, and lower or dependent functions appear in the right portion. Why-How Logic is used to establish these function relationships. The diagram is created by applying the Function Analysis System Technique.

FAST Tree — A method to show function relationships. Higher or more important functions appear at the top of the tree, and lower level functions appear at the bottom of the tree. The same function relationships are maintained that exist in a FAST Diagram. Each tab space to the right indicates one level lower in logic and describes how a particular function is performed or proposed to be performed. One tab space to the left and above a given function tells why that particular function exists or was brought into existence.

Function — As it pertains to function analysis, a function is a name that consists of an active verb and a noun and has three definitions: (1) a task that must be performed which conveys what needs to be accomplished without identifying the method of accomplishment; (2) describes a specific method of how to accomplish its next higher level function; (3) identifies why a lower level function exists and why it must be performed or accomplished.

Function Analysis — The study of functions and the meaning and concepts expressed within the names given to functions by people as they convey their thoughts regarding how they mentally think about a given subject or project. Function names usually consist of an active verb and a measurable noun.

Function Analysis System Technique (FAST) — A technique for analyzing and organizing the functions of systems, products, plans, processes, procedures, facilities, supplies, etc. This technique is designed to stimulate thinking, communication, and creativity and enhances understanding *by organizing the functions into a cause and effect relationship.*

Function Cluster — Three or more functions clustered together which have been organized by asking the Why-How Logic Questions. They are then checked to make sure the logic holds in both directions. Function clusters are also checked for supporting functions by asking the Verification Questions.

Generalizing Functions — A method of changing a function's name so it is more generic and possesses a broader concept and greater potential for stimulating creativity into new areas of thinking.

Higher Level Functions — The functions that appear in the left-hand portion of a FAST Diagram or at the top of a FAST Tree. When other methods of performing these functions are conceived, major advancements become achievable.

Higher Level Logic — The logic used to search for higher level functions. It is accomplished by selecting and inserting the initial basic function into each of the following three questions: (1) Why is it necessary to perform *this function*? (2) What higher level function caused *this function* to come into being? (3) What is really trying to be accomplished when *this function* is performed? When converted to functions, the answers to these questions always move one's thinking to higher logic levels. The purpose of this exercise is to search for the real problem or areas that will yield the greatest creative results.

How Function — Always one logic level below and one tab space to the right of a why function in a FAST Tree and always the next function to the right in a FAST Diagram. More than one how function may exist at the same logic level. A how function describes how a why function is to be accomplished and identifies the method of doing it. However, the method of accomplishing a how function is always identified in the functions listed at the next lower level.

Independent Function — A function whose existence does not depend upon one or more of the other functions being analyzed at a particular moment. In other words, an independent function is a function that exists at a higher logic level or is in a different branch of the FAST Diagram or FAST Tree.

Initial Basic Function — A function that describes the most important task that must be performed from a list of initial functions. All other functions in the list are brought into existence because this function must be performed.

Initial Function — A function name formulated from a statement or fact which describes an action or task that must be performed as a subject, project, or product is first investigated.

Intuitive Logic — The thinking and logic reasoning that take place within a person's mind whenever he or she intuitively plays a role to answer any question.

Logic Diagrams — FAST Diagrams and FAST Trees are considered to be logic diagrams because they logically connect functions together next to each other *in a cause and effect relationship.* A why function *causes* a how function to come into existence, and the *effect* of forming this relationship with the how function is the why function then becomes the reason the how function must be performed. This characteristic makes a logic diagram a compact method of storing a lot of information within a small space.

Logic Holds in Both Directions — This logic relationship pertains to the functions in a function cluster, a FAST Diagram, or a FAST Tree. A how function when inserted into the Why Logic Question must yield the why function. Also, a why function when inserted into the How Logic Question must yield the how function. When these two answers agree, then the logic is said to hold in both directions.

Lower Level Functions — The functions that appear in the right-hand portion of a FAST Diagram and the lower portions of a FAST Tree. They in general deal with the details of how each preceding function is to be accomplished.

Preceding Function — Comes immediately before or to the left of the function being investigated and is always at a higher logic level. It always tells why the function being investigated needs to be performed.

Primary Path Functions — All functions to the right of a basic function in a FAST Diagram or below the basic function in a FAST Tree which comply with the Why-How Logic in both directions are considered to be primary path functions. These functions do not necessarily guarantee reliable performance or provide for all of the acceptability features, but they are considered to be essential if the basic function is to be performed either momentarily or continuously in the manner conceived.

Role-Playing — In FAST Creativity role-playing, a person intuitively or intentionally imagines himself or herself as a molecule, a part, an item, a person, or in a situation where he or she can mentally visualize responding to a question about a function.

Succeeding Function — Comes immediately to the right of the function being investigated. It always tells how this function is to be or could be performed or how it may help to accomplish the function being investigated.

Supporting Function — A function that assists another function to effectively and reliably perform its task. It may be considered to be a function that solves a problem created by the method selected to perform some other function. Supporting functions frequently are required to achieve the required level of reliability and acceptability.

Undisclosing Method — A method of changing a function's name so it does not disclose the existing method described within its name. This has a tendency to broaden a person's understanding of a subject and move his or her thinking into new avenues which otherwise may never be considered.

Why Function — Always one logic level above and one tab space to the left of a how function in a FAST Tree and always the next function to the left of a how function in a FAST Diagram. A why function describes why a how function is necessary to accomplish a task and causes the how function to come into existence.

Why-How Logic — The logic that expands one's understanding and enhances communication between participants. It is a continuous process of asking the Why-How Logic Questions of every function that surfaces as a FAST Diagram or a FAST Tree is being developed. The logic must be pursued until a consensus is reached and the logic holds in both directions.

Why Logic — (1) Used to explore higher level functions in an effort to expand understanding of a given subject. When it is used in this manner, all possible answers are recorded as functions. These answers may be several logic levels above the function being investigated. Higher Level Logic Questions and Why Logic Questions are used to accomplish this. (2) Also used to verify if a how function is a correct response to a How Logic Question for a why function.

THOUGHT-PROVOKING QUESTIONS

Basic Function Determination Question
1. If *this function* didn't have to be performed, would any of the other functions listed still have to be performed?

Higher Level Logic Questions
2. Why is it necessary to perform *this function*?
3. What higher level function caused *this function* to come into being?
4. What is really trying to be accomplished when *this function* is performed?

How Else Logic Question
5. How else can *this function* be accomplished or performed?

How Logic Question
6. How is *this function* actually accomplished or proposed to be accomplished?

This question is asked of a why function and should yield at least one how function as an answer.

Supporting Function Question
    7. When/if *this function* is performed, what other functions must be performed?

This question is asked of all primary path functions to discover if any supporting functions should be added to the logic diagram. Secondary functions also have supporting functions.

Verification Questions
    8. Does *this how function* help *its why function*?
    9. If *this how function* is accomplished, will it help *its why function*?

A "yes" answer to the first question verifies the logic. When in doubt, the second question is asked, which must also yield a "yes" answer to verify the logic. A "no" answer to either of these questions indicates that the how function is a supporting function of the why function.

When/If Logic Question (Same as Supporting Function Question)
    7. When/if *this function* is performed, what other functions must be performed?

Why-How Logic Questions
    10. Why must *this function* be performed?
    6. How is *this function* actually accomplished or proposed to be accomplished?

Why Logic Questions
    2. Why is it necessary to perform *this function*?
    10. Why must *this function* be performed?

This book has free materials available for download from the
Web Added Value™ Resource Center at www.jrosspub.com.

INDEX